24
Topics in Medicinal Chemistry

Aims and Scope

Drug research requires interdisciplinary team-work at the interface between chemistry, biology and medicine. Therefore, the new topic-related series Topics in Medicinal Chemistry will cover all relevant aspects of drug research, e.g. pathobiochemistry of diseases, identification and validation of (emerging) drug targets, structural biology, drugability of targets, drug design approaches, chemogenomics, synthetic chemistry including combinatorial methods, bioorganic chemistry, natural compounds, high-throughput screening, pharmacological in vitro and in vivo investigations, drug-receptor interactions on the molecular level, structure-activity relationships, drug absorption, distribution, metabolism, elimination, toxicology and pharmacogenomics.

In general, special volumes are edited by well known guest editors.

In references Topics in Medicinal Chemistry is abbreviated Top Med Chem and is cited as a journal.

More information about this series at http://www.springer.com/series/7355

Michael S. Wolfe
Editor

Alzheimer's Disease II

With contributions by

C. Ballatore · M. Botta · K.R. Brunden · E.L. Cárdenas · S.R. Choi · A.L. Fallacara · A.K. Ghosh · D.S. Johnson · H.F. Kung · V.M.-Y. Lee · H.L. Osswald · M. Pettersson · K. Ploessl · S. Schenone · A.B. Smith III · I.M.L. Trist · J.Q. Trojanowski · M.S. Wolfe · L. Zhu

Springer

Editor
Michael S. Wolfe
Department of Medicinal Chemistry
University of Kansas
Lawrence, Kansas
USA

ISSN 1862-2461 ISSN 1862-247X (electronic)
Topics in Medicinal Chemistry
ISBN 978-3-319-59459-0 ISBN 978-3-319-59460-6 (eBook)
DOI 10.1007/978-3-319-59460-6

Library of Congress Control Number: 2015955922

Printed on acid-free paper

This Springer imprint is published by Springer Nature
The registered company is Springer International Publishing AG
The registered company address is: Gewerbestrasse 11, 6330 Cham, Switzerland

Preface

Alzheimer's disease (AD) is a devastating neurodegenerative brain disorder that leads to progressive loss of memory and other higher cognitive functions and ultimately death. Age is the greatest risk factor for AD, and therefore longer life spans mean more cases. In the United States alone, over five million people have AD, and worldwide estimates are 20–30 million. These numbers are projected to rise dramatically in the coming decades, creating a crisis that could bankrupt healthcare systems and overwhelm society as a whole.

The need for effective prevention and treatment is urgent, and many laboratories around the world are engaged in research into the root causes and mechanisms of the disease and the discovery of potential therapeutics. Although important gaps remain in our understanding of the molecular basis of AD pathogenesis and progression, considerable progress has been made in this regard.

Two key pathological features are found in the AD brain: amyloid plaques and neurofibrillary tangles. The former is primarily composed of a 42-residue amyloid β-peptide (Aβ), while the latter are composed of the microtubule-associated protein tau. While the deposits themselves may only be disease markers, a large body of evidence, particularly from genetics, strongly suggests that these proteins are directly involved in pathogenesis and progression.

The process apparently begins with the overproduction or under-clearance of the Aβ, a proteolytic product generated through cleavage of the amyloid precursor protein by the sequential action of β- and γ-secretases. Assembly of the amphipathic Aβ peptide in the form of neurotoxic oligomers and other higher-order forms leads to synaptic toxicity and the triggering, through unknown mechanisms, of tau misfolding and aggregation.

This volume first provides an overview of AD biology and the prospects for developing therapeutics (chapter "Targets and Strategies Toward the Development of Alzheimer Therapeutics"). What follows are specific and detailed medicinal chemical strategies for AD drug discovery and the development of diagnostic tools.

Preventing the formation of neurotoxic Aβ through inhibition of β-secretase (chapter "The Design, Development, and Evaluation of BACE1 Inhibitors for the Treatment of Alzheimer's Disease") or modulation of γ-secretase (chapter "γ-Secretase Modulators as Aβ42-Lowering Pharmacological Agents to Treat Alzheimer's Disease") is a leading strategy. As pathological tau is hyperphosphorylated, the responsible kinases are considered important targets (chapter "Inhibitors of Tau-Phosphorylating Kinases"), and because pathological tau is not able to serve its normal function in stabilizing microtubules, small molecules that rescue this loss of function may have therapeutic benefit (chapter "Microtubule-Stabilizing Agents for Alzheimer's and Other Tauopathies"). Finally, early diagnosis of AD is critical for enrolling the right patients into clinical trials and intervening before disease progression has gone too far, and the development of PET imaging agents for amyloid plaques has led to several approved diagnostic agents (chapter "PET Imaging Agents for Alzheimer's Disease").

I would like to thank all the authors who contributed to this volume for their time, thought, and effort in putting together high-quality chapters on these important targets and strategies for AD prevention, treatment, and diagnosis. I also thank all those working on this difficult problem in human health. We all greatly hope that these efforts will speed the day when AD is a scourge of the past.

Lawrence, KS, USA Michael S. Wolfe

Contents

Top Med Chem (2017) 24: 1–26
DOI: 10.1007/7355_2016_14

Published online: 11 December 2016

Targets and Strategies Toward the Development of Alzheimer Therapeutics

Michael S. Wolfe

Abstract Although Alzheimer's disease (AD) is a devastating neurodegenerative disorder affecting tens of millions worldwide, there are no effective disease-modifying therapies and only a handful of symptomatic treatments. Much has been deciphered regarding the molecular basis of AD in the past 25 years, but serious gaps in understanding remain. Aggregation-prone proteins amyloid-β (Aβ) and tau are apparently central to disease pathogenesis and progression; however, the neurotoxic forms of both proteins and the connection between the two remain unclear. Genetic mutations in the precursor protein for Aβ and in a protease that produces Aβ from this precursor cause dominantly inherited early-onset AD. In contrast, an allelic variant of apolipoprotein E is a major risk factor for the much more common late-onset AD, and the encoded protein is critical for Aβ clearance from the brain. Tau is mutated in other forms of dementia, and tau pathology correlates better with neurodegeneration and progression than does Aβ pathology. A variety of therapeutic agents are advancing through the pipeline, many targeting Aβ and tau. New symptomatic treatments are still needed for those who have already progressed too far in the disease process. Advances in drug discovery, new diagnostic tools such as Aβ imaging agents, and improvements in clinical trial design increase the likelihood that effective agents that slow or halt the progression of AD will soon be realized.

Keywords Alzheimer's disease, Amyloid, Neurodegeneration, Tau

---

M.S. Wolfe (✉)
Ann Romney Center for Neurologic Diseases, Brigham and Women's Hospital and Harvard Medical School, Boston, MA, USA
e-mail: mswolfe@partners.org

Contents

Abbreviations

5-HT	5-Hydroxytryptamine or serotonin
AD	Alzheimer's disease
ApoE	Apolipoprotein E
APP	Amyloid-β protein precursor
Aβ	Amyloid-β protein
BACE	β-Site APP-cleaving enzyme
CDK	Cyclin-dependent kinase
CSF	Cerebrospinal fluid
CTF-β or –β	C-Terminal fragment generated by β- or β-secretases
FAD	Familial Alzheimer's disease
FDA	Food and Drug Administration
GSK	Glycogen synthase kinase
MARK	Microtubule affinity-regulating kinase
MCI	Mild cognitive impairment
NMDA	*N*-Methyl-D-aspartate
PET	Positron emission tomography
PS1 or PS2	Presenilin-1 or presenilin-2
SAD	Sporadic Alzheimer's disease
USP	Ubiquitin-specific protease

1 Introduction

Alzheimer's disease (AD) is a devastating neurodegenerative disorder that leads to progressive memory deficiencies and cognitive impairment and ultimately to death within 5–20 years from onset of symptoms. Although many factors such as genetics and lifestyle affect an individual's risk of getting AD, the greatest risk factor by far is age: AD is rare before age 65 and increases dramatically in the years afterward, with a ~40% chance of having AD by age 85. Over 5 million people in the United States and perhaps over 30 million worldwide have AD, with these numbers expected to increase dramatically as demographic changes result in greater numbers of elderly [1].

Much has been learned in the past three decades about the biological basis of the disease, revealing potential means of developing biomarkers for diagnosis, risk assessment, clinical trial enrollment, and drug target engagement, as well as suggesting targets for drug discovery for prevention and treatment. Despite this progress, only a few drugs for AD have been approved, and these are symptomatic treatments that do not affect the underlying progressive neurodegeneration and for this reason soon become ineffective. Indeed, no drugs have been approved for AD since 2003; every clinical candidate brought forward since the approval of the NMDA receptor blocker memantine has failed, and these failures number in the hundreds. The possible reasons for these failures are many, but taken together the inability to identify effective therapeutics suggests insufficient knowledge of disease mechanisms. This overview chapter will summarize the current state of understanding of the biology of the disease along with targets and strategies being pursued to discover effective AD therapeutics, presenting the progress as well as the challenges.

2 AD Presentation and Progression

AD begins with subtle problems with memory, learning, and executive function, termed mild cognitive impairment (MCI). As the disease progresses over the course of years, these problems become more pronounced, with the patient increasingly less able to function without help. Ultimately, family and friends are overwhelmed by the challenges of helping the AD patient meet his or her basic daily needs, preventing the wandering off without notice, and managing the disruptions in the sleep-wake cycle, with professional care in the form of a nursing home being the typical fate for late-stage AD patients. This pattern of disease presentation and progression is the consequence of the relentless loss of synapses and neurons in regions of the brain responsible for learning, memory, and cognition, beginning in the entorhinal cortex, spreading into the hippocampus and the frontal cortex and temporal lobes [2].

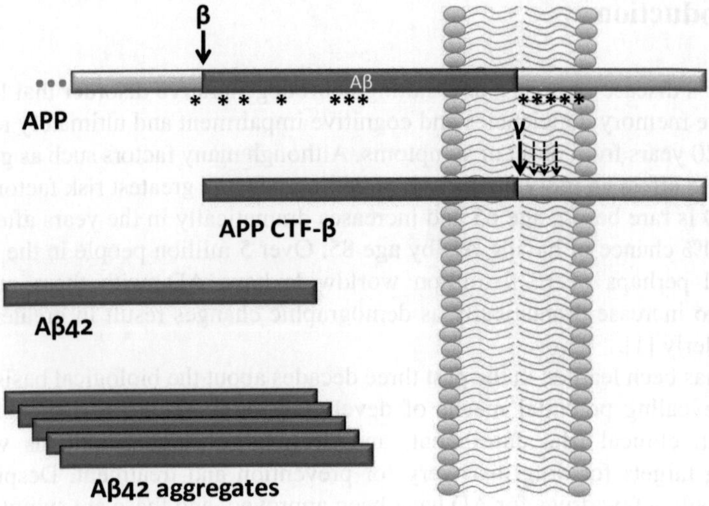

Fig. 1 Proteolytic processing of APP to Aβ, with subsequent aggregation of Aβ42

3 Pathology

Two pathological hallmarks of AD were first described by Alois Alzheimer over a century ago. Silver staining of brain sections taken from a 55-year-old woman who died after developing presenile dementia revealed extra-neuronal deposits of what was dubbed amyloid as well as intraneuronal deposits called neurofibrillary tangles in nerve cell bodies and processes. Decades later, in the 1980s, the extra-neuronal deposits were found to be composed of a 4 kDa peptide called the amyloid-β protein (Aβ) [3], and the intraneuronal deposits were determined to be composed of a 352–441-residue protein called tau [4, 5]. A third pathological feature also became recognized, neuroinflammation, an apparent consequence of over-activation of non-neuronal cells such as microglia and astrocytes [6].

3.1 Amyloid

Aβ is a 38–43-residue secreted peptide and the product of proteolytic processing of the Aβ precursor protein (APP), a type I integral membrane protein of unknown function [7, 8]. The N-terminus of Aβ is formed through cleavage in the lumenal/extracellular region of APP by the membrane-tethered aspartyl protease β-secretase [9], to release the APP ectodomain into the lumenal/extracellular space and leave a 99-residue membrane-bound protein stub (APP CTF-β) (Fig. 1). The C-terminus of Aβ is generated through the action of γ-secretase, an unusual membrane-embedded aspartyl protease complex [10], cutting APP CTF-β within its single transmembrane domain and releasing the APP intracellular domain (AICD) as the other proteolytic product. Alternative processing of APP by α-secretases ADAM-10 and ADAM-17 [11, 12] within the Aβ sequence precludes Aβ production, with the 83-residue protein remnant (APP CTF-α) subsequently cut by γ-secretase to produce an N-terminally truncated form of Aβ called p3. Although the 40-residue form of Aβ (Aβ40) is the primary proteolytic product, the minor 42-residue variant (Aβ42) is much more aggregation prone [13] and is the principle protein found in the cerebral amyloid deposits of AD [14].

Dominantly inherited missense mutations in the gene encoding APP cause familial AD (FAD), with an age of onset before age 60 [15, 16]. These mutations are found in and around the small Aβ region of the much larger APP. Mutations near the β-secretase cleavage site increase proteolysis by the protease, leading to increased production of all forms of Aβ. Mutations within the Aβ region increase the aggregation propensity of Aβ. Mutations within the transmembrane of APP increase the proportion of Aβ42/Aβ40. Thus, all FAD-causing mutations in APP increase the tendency of Aβ to aggregate. Other dominant mutations that cause early-onset FAD are found in the multi-pass membrane proteins presenilin-1 (PS1) [16, 17] and presenilin-2 (PS2) [16, 18], proteins that are the catalytic component of the γ-secretase complex [10, 19]. These mutations also increase the proportion of Aβ42/Aβ40 [10]. Cleavage of APP CTF-β by γ-secretase, however, is more complicated than originally thought. Initial proteolysis at the ε cleavage site near the membrane-cytosolic boundary releases AICD and produces 48- or 49-residue Aβ (Aβ48 or Aβ49), which is sequentially trimmed generally every three amino acids by a carboxypeptidase function of the γ-secretase complex [20, 21]. Thus, there are two pathways to secrete Aβ: Aβ49→Aβ46→Aβ43→Aβ40 and Aβ48→Aβ45→Aβ42→Aβ38, with this last cleavage releasing a tetrapeptide instead of a tripeptide. Most presenilin mutations increase the proportion of ε cleavage giving Aβ48, thereby increasing Aβ42/Aβ40 [22]. These mutations also dramatically reduce the carboxypeptidase function of the γ-secretase complex [23].

Although FAD is clearly associated with altered Aβ production, the much more common late-onset sporadic AD (SAD) is associated with decreased clearance of Aβ [24]. Degradation of Aβ in the brain has been attributed to the metalloprotease neprilysin [25] and insulin-degrading enzyme [26]. Thus, a common feature of AD is increased levels of aggregation-prone forms of Aβ, either through altered production or clearance. For this reason, Aβ has been the major target for AD drug discovery, either directly through antibodies or aggregation inhibitors or indirectly through the secretases or Aβ-degrading enzymes.

The aggregation of Aβ is closely correlated with AD. However, the amyloid plaques do not correlate well spatially with neurodegeneration [2]. This has led to the focus on soluble Aβ oligomers as candidate toxic entities responsible for synaptic and neuronal dysfunction [27]. Synthetic Aβ oligomers as well as natural Aβ oligomers, secreted from cells or extracted from transgenic AD mouse brain or human AD brain, have been shown to affect dendritic spines, synaptic signaling, and electrophysiological responses, including long-term potentiation associated with memory formation [28]. The search for possible receptors for oligomeric Aβ

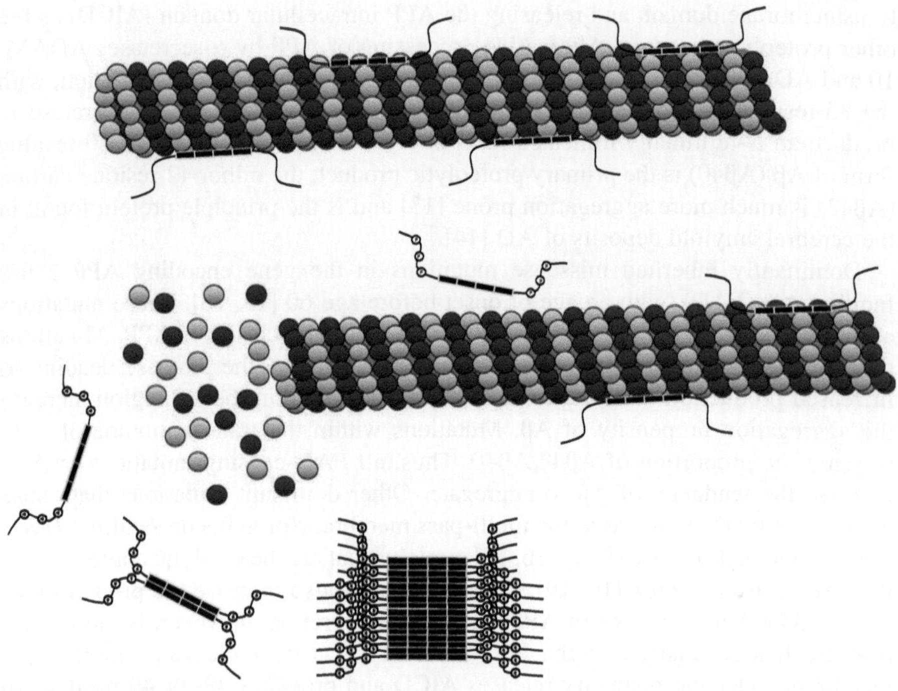

Fig. 2 Destabilization of microtubules upon tau protein release and phosphorylation. Hyperphosphorylation is correlated with tau aggregation and neurofibrillary tangles

has led to the identification of the cellular prion protein as a leading candidate [29], with evidence for downstream effects on NMDA receptors and activation of kinases that phosphorylate tau [30]. Sizes ranging from dimers and trimers [31] to dodecamers [32] and to protofibrils [33] have been suggested as toxic forms of Aβ, but at this time there is no consensus on which of these, if any, is most likely. Indeed, it has been suggested that there is no single toxic species but rather that a toxic "Ab soup" may be responsible [34].

3.2 *Tau*

Tau, the other aggregation-prone protein in AD, is normally associated with microtubules (Fig. 2) [35]. Although tau can help stabilize microtubules, this function is apparently redundant with other proteins, as knockout of tau is not lethal [36]. Crossing tau knockout mice with APP transgenic AD mouse models does not lead to altered Aβ production or deposition but results in attenuation of age-dependent learning and memory deficits [37], suggesting that tau is downstream of Aβ in AD pathogenesis. Dominantly inherited mutations in tau do not cause AD, but they do result in tau pathology, neurodegeneration, and other forms of dementia [38, 39]. These and other findings suggest that Aβ is one of a variety of factors that can trigger pathological tau aggregation and consequent neurodegeneration.

Tau protein in the neurofibrillary tangles of AD is hyperphosphorylated, which has led to efforts to identify the responsible kinases and search for selective inhibitors [40, 41]. Although the assumption is that the increased phosphorylation of tau is important for AD pathogenesis, whether the relationship between tau hyperphosphorylation and neurodegeneration is one of cause and effect or only correlative is unclear. Aggregation of tau in the form of filaments dramatically decreases tau turnover and may therefore simply allow more time for kinases to phosphorylate the protein. Even if hyperphosphorylation is part of the pathogenic process, identifying the specific sites that are most important and the responsible kinases has been fraught with complications, especially as phosphorylation of a given site can affect phosphorylation of another site, and more than one kinase is often capable of phosphorylating a given site [42].

Despite the above critique of the role of hyperphosphorylation in AD pathogenesis, it is important to point out that phosphorylation of tau is critical to regulating its interaction with microtubules [43], and dissociation from microtubules increases the tendency of the protein to aggregate [42] (Fig. 2). Moreover, hyperphosphorylation may play a role in AD pathogenesis due to dissociation of tau and consequent destabilization of microtubules. Although knockout of tau in mice is not lethal and does not lead to neurodegeneration [36], compensation may take place in the context of deletion from conception onward. Thus, it would be important to determine the effects of conditionally knocking out the tau gene in the adult mouse brain. Furthermore, even if loss of microtubule-stabilizing function alone does not cause neurodegeneration, it may be a secondary contributor to the primary insult of tau aggregation. For this reason, microtubule-stabilizing agents have been sought for the potential treatment of AD and other "tauopathies" [44].

The aggregation of tau is also closely linked to AD pathogenesis, with neurofibrillary tangle pathology correlating better with neurodegeneration than amyloid pathology does [2]. Moreover, seeds of aberrant forms of tau are now thought to spread tau dysfunction and pathology from neuron to neuron through synapses, that is, through neuronal networks [45, 46]. This mechanism of transmission from neuron to neuron is reminiscent of how prion diseases manifest themselves in the brain. Indeed, the idea that many proteins that misfold and aggregate in neurodegenerative diseases spread through prion-like means is gaining advocates [47]. If true, this could potentially provide new means of intervening pharmacologically.

3.3 *Neuroinflammation*

The activation of non-neuronal cells, specifically microglial and astrocytes, along with the release of reactive cytokines is also a hallmark of AD pathology [6]. Related to macrophages, microglial are immune cells of the brain, responsible for scavenging for debris, damaged neurons, and infectious agents. Amyloid plaques are among the brain debris engulfed through phagocytosis by microglial. Chronic over-activation of microglial leads to a state called neuroinflammation, which involves release of pro-inflammatory cytokines such as interferon-γ and interleukin-1 and subsequent tau phosphorylation and neuronal dysfunction. Astrocytes are support cells for neurons, providing critical nutrients and even eliciting signaling through release of the neurotransmitter glutamate. Reactive astrocytes are also found near amyloid plaques and can take up and degrade Aβ.

4 Pathogenesis and Risk Factors

The leading model for AD pathogenesis is the amyloid hypothesis [28], which posits that elevation of aggregation-prone forms of Aβ, either through altered production as in FAD or altered clearance as in SAD, results in the formation of toxic Aβ, perhaps soluble oligomers, that disrupt normal synaptic and neuronal function, trigger changes in tau, and elicit neuroinflammation. These effects ultimately lead to neurodegeneration and over years and decades to the clinical manifestation of AD.

The primary risk factor for AD is age. Incidence of AD before age 60 is rare and associated with dominantly inherited FAD mutations in APP and presenilins. After age 60, the risk increases dramatically, so that by age 85 the chance of having AD is ~40–50% [1]. For late-onset SAD, the most important genetic risk factor is apolipoprotein E: a single copy of the E4 allele increases the risk of AD three- to fourfold, and two copies increase the risk 12–15-fold [48]. These findings suggest that ApoE may be a worthwhile target for drug discovery. Diabetes and cardiovascular disease are also important AD risk factors, raising hopes that dietary and lifestyle changes may reduce lifetime risk to AD [49, 50].

5 Animal Models

Transgenic mice have been by far the primary animal models developed for AD. These have been engineered to overexpress FAD mutant forms of human APP and PS1, leading to age-dependent deposition of amyloid plaques and deficiencies in learning and memory [51]. However, these mice have demonstrated neither tau pathology nor frank neurodegeneration, leading to concerns about whether the observed cognitive deficits are elicited through bona fide AD mechanisms and not simply the consequence of gross overexpression of these proteins. To elicit tau pathology, APP/PS1 transgenic mice have been crossed with mice overexpressing human tau with point mutations that cause the AD-related frontotemporal dementia [52]. This cross has no effect on the amyloid production and deposition, but does increase tau pathology and exacerbate cognitive deficits.

Such AD animal models have been used extensively to test candidate AD therapeutics. However, the translation of promising preclinical candidates to the clinic has not been successful, raising serious concerns about the validity of the animal models, particularly of the transgenic mice that highly overexpress aggregation-prone proteins. The concern about artifacts due to overexpression has led to efforts to generate knock-in mice that express disease-causing mutant forms of APP, PS1, or tau under the control of endogenous promoters, with levels and spatiotemporal distribution that is physiological [53–55].

6 Clinical Trials

The design of clinical trials for AD experimental therapeutics is challenging [56]. First, these trials can be quite lengthy, depending on the goal. For symptomatic treatments, 6-month trial is often sufficient, whereas testing for disease modification requires at least 18 months and hundreds of subjects during phase II and thousands of subjects during phase III. This is because the underlying neurodegeneration – the fundamental aspect that is to be modified – is a slow process; extensive time and large numbers of subjects are needed to distinguish a statistically significant effect between treatment and placebo.

Second, identifying those suitable for enrollment requires tests for memory and cognition that are spaced apart in time to determine that the disease is progressive. This is still not sufficient for distinguishing AD from other dementias; biomarkers such as amyloid PET imaging are required for a clearer diagnosis [57]. The use of amyloid imaging agents is particularly critical for testing the efficacy of anti-Aβ therapeutic candidates: the enrollment of non-AD dementia subjects has interfered with the clear interpretation of many previous trials for AD clinical candidates.

Third, biomarkers are needed not only to enroll the right subjects but as a measure of drug effectiveness in engaging its target. In addition to PET imaging for Aβ deposition, CSF Aβ42 is a biomarker [58]. Counterintuitively, CSF levels of this peptide are reduced in AD, presumably due its deposition in the brain. Elevated total tau and phosphorylated tau in the CSF are also considered AD biomarkers. Fourth, the optimal point of intervention is critically important but can be difficult to ascertain. For instance, amyloid pathology can begin 25 years before the onset of symptoms [59], and it is unclear if enrolling presymptomatic subjects 1–2 years prior to the emergence of cognitive deficits will be sufficient for testing anti-Aβ therapeutics.

7 Therapeutic Targets and Strategies

There are many potential points of therapeutic intervention for AD, both for symptomatic and disease-modifying agents. Candidate symptomatic treatments often target a neurotransmitter or its receptors, examples being cholinesterase inhibitors or 5-HT receptor subtype agonists or antagonists. Candidates for disease modification may target Aβ production, aggregation, or clearance; tau aggregation, loss of microtubule-binding function, or interneuronal transmission; and glial cell over-activation.

7.1 *Symptomatic Treatment*

Cholinesterase inhibitors such as donepezil (Aricept) have been a mainstay of AD therapy for nearly 20 years [60]. These agents boost acetylcholine levels at synapses, allowing for stronger cholinergic signaling. This is important because many of the neurons lost in AD critical to learning and memory are cholinergic. Cholinesterase inhibitors, however, do not affect the underlying neurodegeneration and are thus symptomatic treatments. An advantage of these agents is that they boost acetylcholine levels that are naturally released from the presynaptic terminal; that is, they potentiate the signal that is normally produced in time and space in the brain. However, cholinesterase inhibitors typically become ineffective within 6 months to a year, as more and more cholinergic neurons are lost.

Another strategy for cholinergic stimulation is developing muscarinic and nicotinic agonists to target acetylcholine receptor subtypes [61–63]. Although this could lead to chronic stimulation of these receptors, which could result in intolerable side effects, positive allosteric modulators are typically sought, allowing greater sensitivity or responsiveness to the endogenous neurotransmitter. In this way, spatial and temporal control over receptor activation and synaptic firing is maintained. 5-HT6 receptor antagonists have also been considered for symptomatic treatment [64], primarily as an add-on with donepezil. Blocking this serotonin receptor subtype enhances acetylcholine and glutamate function. Modulation of glutamate receptors also provides symptomatic treatment; this is the basis of the approved AD drug memantine, which blocks excess signaling from the NMDA glutamate receptor subtype [60, 65].

7.2 *Targeting Aβ*

Targeting Aβ with small molecules can be indirect, by preventing its production through β- or γ-secretase, or direct, by blocking its aggregation. Many candidates targeting amyloid, directly or indirectly, have been in clinical trials; however,

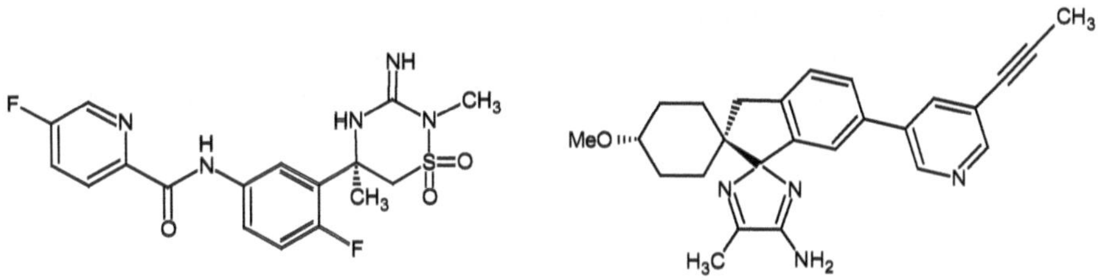

MK8931 AZD3293

Fig. 3 BACE1 inhibitors in late-stage clinical trials

nothing has yet emerged as an approved drug. One reason for this may be what was mentioned earlier, that Aβ deposition begins many years, even decades before the onset of symptoms [59]. Clinical trials have typically enrolled subjects with MCI or early AD. Even a prevention trial, enrolling subjects predicted to show symptoms within the span of the trial, may be too late for an anti-Aβ therapeutic. The process may be tau driven and Aβ independent by this point, so intervening at the level of Aβ accomplishes little. Ongoing prevention trials with agents targeting Aβ are running 5 years, and perhaps there will be some clear benefits of intervening at this stage, several years prior to expected onset of symptoms.

7.2.1 β-Secretase Inhibitors

At present, β-secretase is the top anti-Aβ target for small-molecule therapeutics, with several compounds in the clinic, some in late-stage trials (Fig. 3), including long-term prevention trials [66]. The enzyme β-secretase, also known as β-site APP-cleaving enzyme 1 or BACE1 [67–71], is encoded by a single gene, and knockout of the BACE1 gene in mice eliminates Aβ production in the brain [72–74]. Moreover, crossing BACE1 knockout mice with APP transgenic mice rescues learning and memory deficits [75]. BACE1 knockout mice are viable and fertile; however, they show deficiencies in Schwann cell and oligodendrocyte sheathing of neurons [76] and develop cognitive, emotional, and synaptic problems [77], perhaps due to the inability to cleave the alternative substrate neuregulin-1 [76]. However, this occurs upon knockout of BACE1 from conception. Conditional knockout of the gene in adult mice would provide a better indication of the safety of BACE1 inhibition in elderly AD patients.

The β-secretase enzyme has been crystallized and its high-resolution structure determined, both alone and in complex with small-molecule inhibitors [66, 78]. This has allowed structure-based design and refinement of β-secretase inhibitors with high potency [66]. The biggest obstacle to β-secretase inhibitor development, however, has been brain penetration and retention. Potent inhibitors are often large molecules that do not penetrate the brain readily, and even when they do traverse the blood–brain barrier, they are often ejected by P-glycoprotein

[79]. These obstacles, however, have been largely overcome in recent years, with a number of compounds entering into human trials. Among these clinical candidates, Eli Lilly's LY2811376 could effectively block Aβ production in CSF in healthy volunteers; however, retinal toxicity observed in rodents led to halting human trials [80]. At present, it is unclear if mechanism-based toxicity will be a problem, as β-secretase has a number of other substrates besides APP and neuregulin-1 [81, 82].

7.2.2 γ-Secretase Modulators

The other protease responsible for Aβ production, γ-secretase, has also been a top target for the development of AD therapeutics [83]. However, as mentioned earlier, knockout of components of the γ-secretase complex blocks cleavage of Notch receptors, preventing essential signaling for cell differentiation, resulting in an embryonic lethal phenotype [84, 85]. Notch signaling is also critical for many cell differentiation events in adulthood. Because of this, inhibitors of γ-secretase cause serious side effects such as immunosuppression, skin lesions, and gastrointestinal distress [86–88]. Efforts to titrate the dose of γ-secretase inhibitors to effectively lower CNS Aβ levels while avoiding the toxicities due to deficient Notch signaling failed in advanced human trials, even with γ-secretase inhibitors that purportedly possessed selectivity for blocking proteolysis APP over that of Notch [88, 89].

For this reason, the focus with respect to γ-secretase as a target has shifted to modulators [90]. These are compounds that lower the aggregation-prone Aβ42 selectively, by specifically stimulating the carboxypeptidase trimming of Aβ42 to Aβ38 [91]. At effective concentrations, total Aβ is not lowered, and Notch proteolysis is not affected. These compounds can be very safe and given at high doses [92, 93]. However, potency and brain accessibility have been problems, and at present no γ-secretase modulator is in advanced clinical trials. It should be mentioned that the success of these agents entirely depends on Aβ42 being the pathogenic entity in AD. As mentioned earlier, this matter has not been entirely settled.

7.2.3 Aggregation Inhibitors

The identification of compounds that directly bind to Aβ and Aβ assemblies to prevent aggregation has been a long-studied strategy for the development of AD therapeutics [94]. The challenge of this approach over blocking the secretases is one of stoichiometry, as inhibiting an enzyme can prevent the production of many Aβ molecules, while direct binding requires higher levels of compound, with increased potential for off-target effects. Nevertheless, substoichiometric levels of A-β-binding compounds have been reported to prevent aggregation [95]. A critical issue here is that the compounds should not prevent deposition of Aβ while increasing levels of potentially neurotoxic Aβ oligomers. Other problems with this approach have been demonstrating selectivity and brain penetration. Although

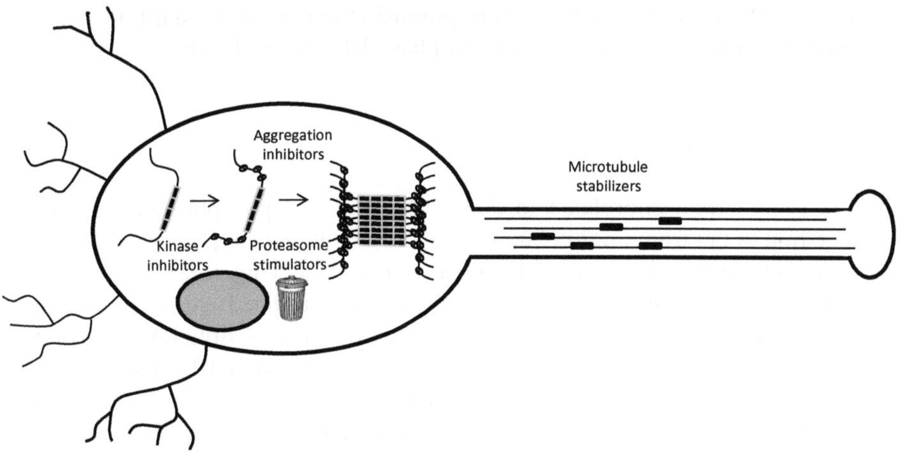

Fig. 4 Strategies for developing therapeutics targeting tau

many compounds have been reported to inhibit Aβ aggregation, at present none are in advanced human trials.

7.3 Targeting Tau

The deposition of tau in the form of filaments found in neurofibrillary tangles is another canonical feature of AD pathology. As mentioned earlier, tau is critical for the development of learning and memory deficits in APP transgenic mice [37]. This and other evidence suggest that tau is downstream of Aβ in the AD pathogenic process. As with Aβ, however, the pathogenic form of tau is not known, which makes it challenging to develop a strategy for developing effective therapeutics that target this protein. Nevertheless, several approaches have been pursued over the years, described below (Fig. 4).

7.3.1 Aggregation Inhibitors

Although the toxic form of tau is not known, its aggregation is closely coupled to pathogenesis, not only for AD but for other neurodegenerative diseases collectively called "tauopathies" [96]. As with Aβ aggregation inhibitors, the search for tau aggregation inhibitors faces challenges of identifying compounds that are potent, selective, and accessible to the brain. Nevertheless, the dye methylene blue appears to have these properties [97], although the compound has also been reported to have other effects such as enhancing protein degradation through the proteasome and autophagy and regulating gene expression [98–100]. In vivo, methylene blue can prevent tau aggregation as well as cognitive deficits in tau transgenic mice

[101, 102]. Phase II clinical trials were promising but controversial [103], and the compound failed in a recently concluded phase III trial [104, 105].

7.3.2 Kinase Inhibitors

As tau is hyperphosphorylated in AD [106, 107] and phosphorylation leads to dissociation of tau from microtubules and its aggregation [108], considerable effort has gone into identifying responsible kinases and identifying potent, selective, and brain-penetrating inhibitors of these kinases. As mentioned earlier, tau contains many phosphorylation sites, and which sites might be responsible for tau toxicity is unclear [41]. Moreover, a given site may be phosphorylated by more than one kinase, and a given kinase has other substrates besides tau. Nevertheless, several kinases have been particularly implicated in the pathological phosphorylation of tau, including GSK-3β, CDK5, and MARKs. The development of effective inhibitors has also been reported, including examples of efficacy in transgenic mice [109]. Despite these findings, there are concerns about on-target toxicity, especially with GSK-3β, which is involved in glucose utilization. At present, no inhibitors of kinases targeting tau are in advanced clinical trials.

7.3.3 Microtubule Stabilizers

Phosphorylation of tau regulates its interaction with microtubules, affecting the role of tau in microtubule stabilization and dynamics [43, 110]. As discussed earlier, the aggregation of tau may be the primary neurotoxic pathway for this protein, but loss of function with respect to microtubules may also be a contributing factor [111]. For this reason, microtubule-stabilizing agents have been sought for the potential treatment of AD and other tauopathies [44]. Such agents have been used as anticancer therapeutics, as microtubules are also critical for mitosis and cell replication. This fact raises the potential concern for microtubule stabilizers: in addition to requiring potency, selectivity, and brain penetration, these compounds must avoid toxic consequences of blocking normal cell replication with chronic administration. Nevertheless, promising compounds have emerged, with one (TPI-287) [112] presently in clinical trials for AD as well as two other tauopathies.

7.3.4 Increasing Clearance

Facilitating the proteolytic breakdown of tau via the proteasome is another strategy for preventing the pathological intraneuronal aggregation [113]. Although stimulation of enzymes is typically challenging, the proteasome has multiple subunits and associated regulatory proteins that open up possibilities for increasing the degradation of misfolded proteins. One example is inhibition of the proteasome component USP14 [114], which normally trims ubiquitin chains from proteins poised for proteasomal degradation. Thus USP14 inhibits protein degradation, and inhibiting this ubiquitin hydrolase with a small molecule could enhance the intracellular clearance of tau protein. However, in vivo deficiency of USP14 revealed that this enzyme is essential for proper synaptic transmission and development of neuromuscular junctions [115]. In addition, USP14 knockout mice showed no decrease in total tau, but a concerning increase in phosphorylated tau [116]. Thus, USP14 may not be a worthwhile target, although the general idea of stimulating proteasomal degradation of tau still has merit.

7.4 *Targeting ApoE*

Targeting ApoE has been more challenging than targeting tau or Aβ. ApoE is apparently involved in Aβ clearance; however, it remains unresolved whether up- or downregulation of ApoE would be beneficial. Thus, even though transcription factors such as retinoic acid receptors control ApoE gene expression, it remains unclear whether agonists or antagonists to these receptors would be therapeutic or whether activating or blocking these receptors would be safe. Another approach has been to identify compounds that alter the conformation of the AD-associated ApoE4 protein so that its conformation is closer to those of ApoE3 or ApoE2 [117].

8 Imaging Agents

Critical to proper diagnosis, clinical trial selection, drug target engagement, and efficacy studies are the identification of suitable biomarkers. Toward this end, the development of imaging agents has been the one bright spot for medicinal chemistry for AD, with several agents approved for human use [118, 119]. At present all the approved imaging agents (Fig. 5) target amyloid deposits, although tau imaging agents are in the pipeline. The pharmacokinetic requirements for imaging agents are different than for therapeutic agents; for the former, high brain uptake and fast washout from normal brain are optimal.

8.1 *Amyloid*

Several positron emission tomography (PET) imaging agents for amyloid deposition have been approved, although only as a way to rule out AD in cases where little or no plaques are present [118, 119]. These agents can support a diagnosis of AD but cannot be used to diagnose AD on their own. Still, this is an important advance, especially for enrolling subjects into clinical trials for candidate AD therapeutics, as one in five patients diagnosed with AD is later found not to have the disease

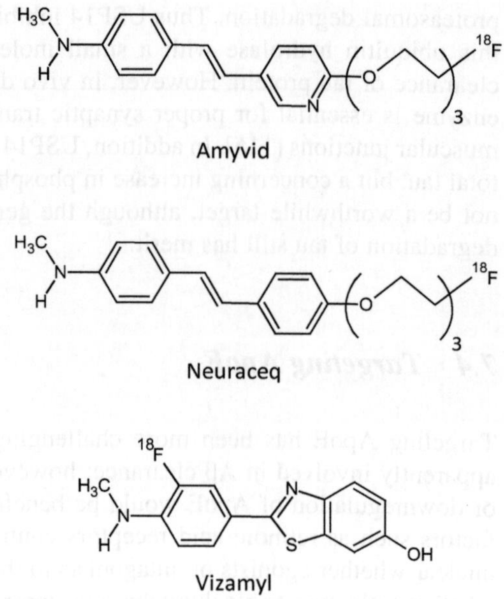

Fig. 5 Approved amyloid imaging agents

[120]. This is because other forms of dementia (e.g., vascular dementia, frontotemporal dementia) can mimic and overlap with AD clinical presentation. The inadvertent inclusion of non-AD subjects has confounded efforts to determine efficacy of AD clinical candidates. These imaging agents have also proven useful for demonstrating target engagement by experimental therapeutics, for instance, with an anti-Aβ antibody that was shown to clear amyloid plaques [121]. The approved imaging agents use ^{18}F as the positron emitter, as this isotope has a 110 min half-life, compared with a 20 min half-life for ^{11}C.

8.2 *Tau*

No imaging agents for tau have been approved yet, although a number of PET imaging agents are in development and under clinical investigation [119]. For use in AD diagnosis, these agents require high specificity for tau aggregates over amyloid deposits, as the latter are much more predominant than the former. However, even without high specificity, a tau imaging agent might be useful in the diagnosis of other tauopathies, especially if amyloid deposition has been ruled out with an Aβ imaging agent. As tau pathology spreads from one region of the brain to another in a fairly predictable course in AD [2] and is more closely correlated with neurodegeneration, both spatially and temporally, than Aβ pathology is, tau imaging agents may ultimately be more useful for diagnosis and staging of the disease as well as for tracking target engagement and efficacy for anti-tau clinical candidates.

9 Perspective

The understanding of the molecular basis of AD has advanced considerably in the past 25 years. The discovery of genes that cause FAD clearly point to some form of Aβ as pathogenic. Studies in mice strongly suggest that pathogenic tau lies downstream of Aβ. However, major gaps remain, including definitive identification of the specific pathogenic forms of Aβ and tau, the connection between these two aggregation-prone proteins, and the role of risk factor ApoE. Transgenic mice mimic some aspects of AD pathology and presentation; however, positive results with numerous experimental therapeutics in these animals have not translated into any approved AD drugs for human use.

Aβ remains the top target for small molecules in the pipeline, particularly through inhibition of β-secretase. However, Aβ deposition can begin over two decades before symptoms, raising concerns that the pathogenic process may become tau driven and Aβ independent many years before anti-Aβ agents can be practically given. For this reason, interest in targeting tau is on the rise, with multiple approaches in development. The further development of symptomatic treatments remains important for those who have already progressed to mild-to-moderate AD and may be beyond the reach of disease-modifying therapies. Finally, imaging agents have been a success story for medicinal chemistry, with several approved agents that detect Aβ deposition in the brain. These imaging agents are already helping to enroll the right subjects for clinical trials and demonstrate target engagement with anti-Aβ clinical candidates. Better mechanistic understanding, expansion of targets and approaches to drug development, validation of more biomarkers, and refinements in the design of clinical trials should together increase the likelihood that effective therapeutics for AD will soon be realized.

Acknowledgments I thank all the contributing authors to this special issue on important approaches to the diagnosis and treatment of AD.

References

1. Alzheimer's Association (2015) 2015 Alzheimer's disease facts and figures. Alzheimers Dement 11(3):332–384
2. Braak H, Braak E (1991) Neuropathological stageing of Alzheimer-related changes. Acta Neuropathol 82(4):239–259
3. Glenner GG, Wong CW (1984) Alzheimer's disease: initial report of the purification and characterization of a novel cerebrovascular amyloid protein. Biochem Biophys Res Commun 120:885–890
4. Grundke-Iqbal I, Iqbal K, Quinlan M, Tung YC, Zaidi MS, Wisniewski HM (1986) Microtubule-associated protein tau. A component of Alzheimer paired helical filaments. J Biol Chem 261(13):6084–6089
5. Kosik KS, Joachim CL, Selkoe DJ (1986) Microtubule-associated protein tau (tau) is a major antigenic component of paired helical filaments in Alzheimer disease. Proc Natl Acad Sci U S A 83(11):4044–4048

6. Heneka MT, Carson MJ, El Khoury J, Landreth GE, Brosseron F, Feinstein DL, Jacobs AH, Wyss-Coray T, Vitorica J, Ransohoff RM, Herrup K, Frautschy SA, Finsen B, Brown GC, Verkhratsky A, Yamanaka K, Koistinaho J, Latz E, Halle A, Petzold GC, Town T, Morgan D, Shinohara ML, Perry VH, Holmes C, Bazan NG, Brooks DJ, Hunot S, Joseph B, Deigendesch N, Garaschuk O, Boddeke E, Dinarello CA, Breitner JC, Cole GM, Golenbock DT, Kummer MP (2015) Neuroinflammation in Alzheimer's disease. Lancet Neurol 14 (4):388–405. doi:10.1016/s1474-4422(15)70016-5
7. Tanzi RE, Gusella JF, Watkins PC, Bruns GA, St George-Hyslop P, Van Keuren ML, Patterson D, Pagan S, Kurnit DM, Neve RL (1987) Amyloid beta protein gene: cDNA, mRNA distribution, and genetic linkage near the Alzheimer locus. Science 235 (4791):880–884
8. Dyrks T, Weidemann A, Multhaup G, Salbaum JM, Lemaire HG, Kang J, Muller-Hill B, Masters CL, Beyreuther K (1988) Identification, transmembrane orientation and biogenesis of the amyloid A4 precursor of Alzheimer's disease. EMBO J 7(4):949–957
9. Cole SL, Vassar R (2008) The role of APP processing by BACE1, the beta-secretase, in Alzheimer's disease pathophysiology. J Biol Chem 283:29621–29625
10. Wolfe MS (2006) The gamma-secretase complex: membrane-embedded proteolytic ensemble. Biochemistry 45(26):7931–7939
11. Vincent B, Paitel E, Saftig P, Frobert Y, Hartmann D, De Strooper B, Grassi J, Lopez-Perez E, Checler F (2001) The disintegrins ADAM10 and TACE contribute to the constitutive and phorbol ester-regulated normal cleavage of the cellular prion protein. J Biol Chem 276 (41):37743–37746
12. Buxbaum JD, Liu KN, Luo Y, Slack JL, Stocking KL, Peschon JJ, Johnson RS, Castner BJ, Cerretti DP, Black RA (1998) Evidence that tumor necrosis factor α converting enzyme is involved in regulated α-secretase cleavage of the Alzheimer's amyloid protein precursor. J Biol Chem 273:27765–27767
13. Jarrett JT, Berger EP, Lansbury PT Jr (1993) The carboxy terminus of the beta amyloid protein is critical for the seeding of amyloid formation: implications for the pathogenesis of Alzheimer's disease. Biochemistry 32(18):4693–4697
14. Iwatsubo T, Odaka A, Suzuki N, Mizusawa H, Nukina N, Ihara Y (1994) Visualization of A beta 42(43) and A beta 40 in senile plaques with end-specific A beta monoclonals: evidence that an initially deposited species is A beta 42(43). Neuron 13(1):45–53
15. Chartier-Harlin MC, Crawford F, Houlden H, Warren A, Hughes D, Fidani L, Goate A, Rossor M, Roques P, Hardy J et al (1991) Early-onset Alzheimer's disease caused by mutations at codon 717 of the beta-amyloid precursor protein gene. Nature 353 (6347):844–846
16. Tanzi RE, Bertram L (2005) Twenty years of the Alzheimer's disease amyloid hypothesis: a genetic perspective. Cell 120(4):545–555
17. Sherrington R, Rogaev EI, Liang Y, Rogaeva EA, Levesque G, Ikeda M, Chi H, Lin C, Li G, Holman K, Tsuda T, Mar L, Foncin JF, Bruni AC, Montesi MP, Sorbi S, Rainero I, Pinessi L, Nee L, Chumakov I, Pollen D, Brookes A, Sanseau P, Polinsky RJ, Wasco W, Da Silva HA, Haines JL, Perkicak-Vance MA, Tanzi RE, Roses AD, Fraser PE, Rommens JM, St George-Hyslop PH (1995) Cloning of a gene bearing missense mutations in early-onset familial Alzheimer's disease. Nature 375(6534):754–760
18. Levy-Lahad E, Wasco W, Poorkaj P, Romano DM, Oshima J, Pettingell WH, Yu CE, Jondro PD, Schmidt SD, Wang K, Crowley AC, Fu YH, Guenette SY, Galas D, Nemens E, Wijsman EM, Bird TD, Schellenberg GD, Tanzi RE (1995) Candidate gene for the chromosome 1 familial Alzheimer's disease locus. Science 269(5226):973–977
19. Wolfe MS, Xia W, Ostaszewski BL, Diehl TS, Kimberly WT, Selkoe DJ (1999) Two transmembrane aspartates in presenilin-1 required for presenilin endoproteolysis and γ-secretase activity. Nature 398:513–517
20. Wolfe MS (2012) Processive proteolysis by gamma-secretase and the mechanism of Alzheimer's disease. Biol Chem 393(9):899–905

21. Takami M, Nagashima Y, Sano Y, Ishihara S, Morishima-Kawashima M, Funamoto S, Ihara Y (2009) gamma-Secretase: successive tripeptide and tetrapeptide release from the transmembrane domain of beta-carboxyl terminal fragment. J Neurosci 29(41):13042–13052
22. Sato T, Dohmae N, Qi Y, Kakuda N, Misonou H, Mitsumori R, Maruyama H, Koo EH, Haass C, Takio K, Morishima-Kawashima M, Ishiura S, Ihara Y (2003) Potential link between amyloid beta-protein 42 and C-terminal fragment gamma 49-99 of beta-amyloid precursor protein. J Biol Chem 278(27):24294–24301
23. Fernandez MA, Klutkowski JA, Freret T, Wolfe MS (2014) Alzheimer presenilin-1 mutations dramatically reduce trimming of long amyloid beta-peptides (Abeta) by gamma-secretase to increase 42-to-40-residue Abeta. J Biol Chem 289(45):31043–31052
24. Mawuenyega KG, Sigurdson W, Ovod V, Munsell L, Kasten T, Morris JC, Yarasheski KE, Bateman RJ (2010) Decreased clearance of CNS beta-amyloid in Alzheimer's disease. Science 330(6012):1774. doi:10.1126/science.1197623
25. Iwata N, Tsubuki S, Takaki Y, Watanabe K, Sekiguchi M, Hosoki E, Kawashima-Morishima-M, Lee HJ, Hama E, Sekine-Aizawa Y, Saido TC (2000) Identification of the major Abeta1-42-degrading catabolic pathway in brain parenchyma: suppression leads to biochemical and pathological deposition. Nat Med 6(2):143–150
26. Farris W, Mansourian S, Chang Y, Lindsley L, Eckman EA, Frosch MP, Eckman CB, Tanzi RE, Selkoe DJ, Guenette S (2003) Insulin-degrading enzyme regulates the levels of insulin, amyloid beta-protein, and the beta-amyloid precursor protein intracellular domain in vivo. Proc Natl Acad Sci U S A 100(7):4162–4167
27. Walsh DM, Klyubin I, Fadeeva JV, Cullen WK, Anwyl R, Wolfe MS, Rowan MJ, Selkoe DJ (2002) Naturally secreted oligomers of amyloid beta protein potently inhibit hippocampal long-term potentiation in vivo. Nature 416(6880):535–539
28. Selkoe DJ, Hardy J (2016) The amyloid hypothesis of Alzheimer's disease at 25 years. EMBO Mol Med. doi:10.15252/emmm.201606210
29. Lauren J, Gimbel DA, Nygaard HB, Gilbert JW, Strittmatter SM (2009) Cellular prion protein mediates impairment of synaptic plasticity by amyloid-beta oligomers. Nature 457 (7233):1128–1132
30. Um JW, Nygaard HB, Heiss JK, Kostylev MA, Stagi M, Vortmeyer A, Wisniewski T, Gunther EC, Strittmatter SM (2012) Alzheimer amyloid-beta oligomer bound to postsynaptic prion protein activates Fyn to impair neurons. Nat Neurosci 15(9):1227–1235. doi:10.1038/nn.3178
31. Shankar GM, Li S, Mehta TH, Garcia-Munoz A, Shepardson NE, Smith I, Brett FM, Farrell MA, Rowan MJ, Lemere CA, Regan CM, Walsh DM, Sabatini BL, Selkoe DJ (2008) Amyloid-beta protein dimers isolated directly from Alzheimer's brains impair synaptic plasticity and memory. Nat Med 14(8):837–842
32. Lesne S, Koh MT, Kotilinek L, Kayed R, Glabe CG, Yang A, Gallagher M, Ashe KH (2006) A specific amyloid-beta protein assembly in the brain impairs memory. Nature 440 (7082):352–357
33. Walsh DM, Hartley DM, Kusumoto Y, Fezoui Y, Condron MM, Lomakin A, Benedek GB, Selkoe DJ, Teplow DB (1999) Amyloid beta-protein fibrillogenesis. Structure and biological activity of protofibrillar intermediates. J Biol Chem 274(36):25945–25952
34. Benilova I, Karran E, De Strooper B (2012) The toxic Abeta oligomer and Alzheimer's disease: an emperor in need of clothes. Nat Neurosci 15(3):349–357
35. Weingarten MD, Lockwood AH, Hwo SY, Kirschner MW (1975) A protein factor essential for microtubule assembly. Proc Natl Acad Sci U S A 72(5):1858–1862
36. Ikegami S, Harada A, Hirokawa N (2000) Muscle weakness, hyperactivity, and impairment in fear conditioning in tau-deficient mice. Neurosci Lett 279(3):129–132
37. Roberson ED, Scearce-Levie K, Palop JJ, Yan F, Cheng IH, Wu T, Gerstein H, Yu GQ, Mucke L (2007) Reducing endogenous tau ameliorates amyloid beta-induced deficits in an Alzheimer's disease mouse model. Science 316(5825):750–754

38. Hutton M, Lendon CL, Rizzu P, Baker M, Froelich S, Houlden H, Pickering-Brown S, Chakraverty S, Isaacs A, Grover A, Hackett J, Adamson J, Lincoln S, Dickson D, Davies P, Petersen RC, Stevens M, de Graaff E, Wauters E, van Baren J, Hillebrand M, Joosse M, Kwon JM, Nowotny P, Heutink P et al (1998) Association of missense and 5'-splice-site mutations in tau with the inherited dementia FTDP-17. Nature 393(6686):702–705
39. Wolfe MS (2009) Tau mutations in neurodegenerative diseases. J Biol Chem 284 (10):6021–6025
40. Johnson GV, Stoothoff WH (2004) Tau phosphorylation in neuronal cell function and dysfunction. J Cell Sci 117(Pt 24):5721–5729
41. Tell V, Hilgeroth A (2013) Recent developments of protein kinase inhibitors as potential AD therapeutics. Front Cell Neurosci 7:189
42. Hanger DP, Anderton BH, Noble W (2009) Tau phosphorylation: the therapeutic challenge for neurodegenerative disease. Trends Mol Med 15(3):112–119. doi:10.1016/j.molmed.2009.01.003
43. Lindwall G, Cole RD (1984) Phosphorylation affects the ability of tau protein to promote microtubule assembly. J Biol Chem 259(8):5301–5305
44. Ballatore C, Brunden KR, Huryn DM, Trojanowski JQ, Lee VM, Smith AB 3rd (2012) Microtubule stabilizing agents as potential treatment for Alzheimer's disease and related neurodegenerative tauopathies. J Med Chem 55(21):8979–8996. doi:10.1021/jm301079z
45. de Calignon A, Polydoro M, Suarez-Calvet M, William C, Adamowicz DH, Kopeikina KJ, Pitstick R, Sahara N, Ashe KH, Carlson GA, Spires-Jones TL, Hyman BT (2012) Propagation of tau pathology in a model of early Alzheimer's disease. Neuron 73(4):685–697
46. Liu L, Drouet V, Wu JW, Witter MP, Small SA, Clelland C, Duff K (2012) Trans-synaptic spread of tau pathology in vivo. PLoS One 7(2), e31302
47. Jucker M, Walker LC (2013) Self-propagation of pathogenic protein aggregates in neurodegenerative diseases. Nature 501(7465):45–51. doi:10.1038/nature12481
48. Corder EH, Saunders AM, Strittmatter WJ, Schmechel DE, Gaskell PC, Small GW, Roses AD, Haines JL, Pericak-Vance MA (1993) Gene dose of apolipoprotein E type 4 allele and the risk of Alzheimer's disease in late onset families. Science 261(5123):921–923
49. Biessels GJ, Strachan MW, Visseren FL, Kappelle LJ, Whitmer RA (2014) Dementia and cognitive decline in type 2 diabetes and prediabetic stages: towards targeted interventions. Lancet Diabetes Endocrinol 2(3):246–255. doi:10.1016/s2213-8587(13)70088-3
50. Norton S, Matthews FE, Barnes DE, Yaffe K, Brayne C (2014) Potential for primary prevention of Alzheimer's disease: an analysis of population-based data. Lancet Neurol 13 (8):788–794. doi:10.1016/s1474-4422(14)70136-x
51. Ashe KH, Zahs KR (2010) Probing the biology of Alzheimer's disease in mice. Neuron 66 (5):631–645
52. Oddo S, Caccamo A, Shepherd JD, Murphy MP, Golde TE, Kayed R, Metherate R, Mattson MP, Akbari Y, LaFerla FM (2003) Triple-transgenic model of Alzheimer's disease with plaques and tangles: intracellular Abeta and synaptic dysfunction. Neuron 39(3):409–421
53. Saito T, Matsuba Y, Mihira N, Takano J, Nilsson P, Itohara S, Iwata N, Saido TC (2014) Single App knock-in mouse models of Alzheimer's disease. Nat Neurosci 17(5):661–663. doi:10.1038/nn.3697
54. Vidal R, Sammeta N, Garringer HJ, Sambamurti K, Miravalle L, Lamb BT, Ghetti B (2012) The Psen1-L166P-knock-in mutation leads to amyloid deposition in human wild-type amyloid precursor protein YAC transgenic mice. FASEB J 26(7):2899–2910. doi:10.1096/fj.12-205542
55. Gilley J, Seereeram A, Ando K, Mosely S, Andrews S, Kerschensteiner M, Misgeld T, Brion JP, Anderton B, Hanger DP, Coleman MP (2012) Age-dependent axonal transport and locomotor changes and tau hypophosphorylation in a "P301L" tau knockin mouse. Neurobiol Aging 33(3):621.e1–621.e15. doi:10.1016/j.neurobiolaging.2011.02.014
56. Schneider LS, Mangialasche F, Andreasen N, Feldman H, Giacobini E, Jones R, Mantua V, Mecocci P, Pani L, Winblad B, Kivipelto M (2014) Clinical trials and late-stage drug development for Alzheimer's disease: an appraisal from 1984 to 2014. J Intern Med 275 (3):251–283. doi:10.1111/joim.12191

57. Morris E, Chalkidou A, Hammers A, Peacock J, Summers J, Keevil S (2016) Diagnostic accuracy of 18F amyloid PET tracers for the diagnosis of Alzheimer's disease: a systematic review and meta-analysis. Eur J Nucl Med Mol Imaging 43(2):374–385. doi:10.1007/s00259-015-3228-x
58. Blennow K, Mattsson N, Scholl M, Hansson O, Zetterberg H (2015) Amyloid biomarkers in Alzheimer's disease. Trends Pharmacol Sci 36(5):297–309. doi:10.1016/j.tips.2015.03.002
59. Bateman RJ, Xiong C, Benzinger TL, Fagan AM, Goate A, Fox NC, Marcus DS, Cairns NJ, Xie X, Blazey TM, Holtzman DM, Santacruz A, Buckles V, Oliver A, Moulder K, Aisen PS, Ghetti B, Klunk WE, McDade E, Martins RN, Masters CL, Mayeux R, Ringman JM, Rossor MN, Schofield PR, Sperling RA, Salloway S, Morris JC (2012) Clinical and biomarker changes in dominantly inherited Alzheimer's disease. N Engl J Med 367(9):795–804. doi:10.1056/NEJMoa1202753
60. Wilcock GK (2008) The pharmacological treatment of Alzheimer's disease with cholinesterase inhibitors and memantine. In: Cuello AC (ed) Pharmacological mechanisms in Alzheimer therapeutics. Springer, New York, pp 36–49
61. Melancon BJ, Tarr JC, Panarese JD, Wood MR, Lindsley CW (2013) Allosteric modulation of the M1 muscarinic acetylcholine receptor: improving cognition and a potential treatment for schizophrenia and Alzheimer's disease. Drug Discov Today 18(23–24):1185–1199. doi:10.1016/j.drudis.2013.09.005
62. Lombardo S, Maskos U (2015) Role of the nicotinic acetylcholine receptor in Alzheimer's disease pathology and treatment. Neuropharmacology 96(Pt B):255–262. doi:10.1016/j.neuropharm.2014.11.018
63. Valles AS, Borroni MV, Barrantes FJ (2014) Targeting brain alpha7 nicotinic acetylcholine receptors in Alzheimer's disease: rationale and current status. CNS Drugs 28(11):975–987. doi:10.1007/s40263-014-0201-3
64. Upton N, Chuang TT, Hunter AJ, Virley DJ (2008) 5-HT6 receptor antagonists as novel cognitive enhancing agents for Alzheimer's disease. Neurotherapeutics 5(3):458–469. doi:10.1016/j.nurt.2008.05.008
65. Lipton SA (2005) The molecular basis of memantine action in Alzheimer's disease and other neurologic disorders: low-affinity, uncompetitive antagonism. Curr Alzheimer Res 2(2):155–165
66. Ghosh AK, Osswald HL (2014) BACE1 (beta-secretase) inhibitors for the treatment of Alzheimer's disease. Chem Soc Rev 43(19):6765–6813. doi:10.1039/c3cs60460h
67. Vassar R, Bennett BD, Babu-Khan S, Kahn S, Mendiaz EA, Denis P, Teplow DB, Ross S, Amarante P, Loeloff R, Luo Y, Fisher S, Fuller J, Edenson S, Lile J, Jarosinski MA, Biere AL, Curran E, Burgess T, Louis JC, Collins F, Treanor J, Rogers G, Citron M (1999) beta-Secretase cleavage of Alzheimer's amyloid precursor protein by the transmembrane aspartic protease BACE. Science 286(5440):735–741
68. Sinha S, Anderson JP, Barbour R, Basi GS, Caccavello R, Davis D, Doan M, Dovey HF, Frigon N, Hong J, Jacobson-Croak K, Jewett N, Keim P, Knops J, Lieberburg I, Power M, Tan H, Tatsuno G, Tung J, Schenk D, Seubert P, Suomensaari SM, Wang S, Walker D, John V (1999) Purification and cloning of amyloid precursor protein beta-secretase from human brain. Nature 402(6761):537–540
69. Yan R, Bienkowski MJ, Shuck ME, Miao H, Tory MC, Pauley AM, Brashier JR, Stratman NC, Mathews WR, Buhl AE, Carter DB, Tomasselli AG, Parodi LA, Heinrikson RL, Gurney ME (1999) Membrane-anchored aspartyl protease with Alzheimer's disease beta-secretase activity. Nature 402(6761):533–537
70. Hussain I, Powell D, Howlett DR, Tew DG, Meek TD, Chapman C, Gloger IS, Murphy KE, Southan CD, Ryan DM, Smith TS, Simmons DL, Walsh FS, Dingwall C, Christie G (1999) Identification of a novel aspartic protease (Asp 2) as beta-secretase. Mol Cell Neurosci 14 (6):419–427
71. Lin X, Koelsch G, Wu S, Downs D, Dashti A, Tang J (2000) Human aspartic protease memapsin 2 cleaves the beta-secretase site of beta-amyloid precursor protein. Proc Natl Acad Sci U S A 97(4):1456–1460

72. Luo Y, Bolon B, Kahn S, Bennett BD, Babu-Khan S, Denis P, Fan W, Kha H, Zhang J, Gong Y, Martin L, Louis JC, Yan Q, Richards WG, Citron M, Vassar R (2001) Mice deficient in BACE1, the Alzheimer's beta-secretase, have normal phenotype and abolished beta-amyloid generation. Nat Neurosci 4(3):231–232
73. Cai H, Wang Y, McCarthy D, Wen H, Borchelt DR, Price DL, Wong PC (2001) BACE1 is the major beta-secretase for generation of Abeta peptides by neurons. Nat Neurosci 4 (3):233–234
74. Roberds SL, Anderson J, Basi G, Bienkowski MJ, Branstetter DG, Chen KS, Freedman SB, Frigon NL, Games D, Hu K, Johnson-Wood K, Kappenman KE, Kawabe TT, Kola I, Kuehn R, Lee M, Liu W, Motter R, Nichols NF, Power M, Robertson DW, Schenk D, Schoor M, Shopp GM, Shuck ME, Sinha S, Svensson KA, Tatsuno G, Tintrup H, Wijsman J, Wright S, McConlogue L (2001) BACE knockout mice are healthy despite lacking the primary beta-secretase activity in brain: implications for Alzheimer's disease therapeutics. Hum Mol Genet 10(12):1317–1324
75. Ohno M, Sametsky EA, Younkin LH, Oakley H, Younkin SG, Citron M, Vassar R, Disterhoft JF (2004) BACE1 deficiency rescues memory deficits and cholinergic dysfunction in a mouse model of Alzheimer's disease. Neuron 41(1):27–33
76. Willem M, Garratt AN, Novak B, Citron M, Kaufmann S, Rittger A, DeStrooper B, Saftig P, Birchmeier C, Haass C (2006) Control of peripheral nerve myelination by the beta-secretase BACE1. Science 314(5799):664–666
77. Savonenko AV, Melnikova T, Laird FM, Stewart KA, Price DL, Wong PC (2008) Alteration of BACE1-dependent NRG1/ErbB4 signaling and schizophrenia-like phenotypes in BACE1-null mice. Proc Natl Acad Sci U S A 105(14):5585–5590
78. Hong L, Koelsch G, Lin X, Wu S, Terzyan S, Ghosh AK, Zhang XC, Tang J (2000) Structure of the protease domain of memapsin 2 (beta-secretase) complexed with inhibitor. Science 290 (5489):150–153
79. Meredith JE Jr, Thompson LA, Toyn JH, Marcin L, Barten DM, Marcinkeviciene J, Kopcho L, Kim Y, Lin A, Guss V, Burton C, Iben L, Polson C, Cantone J, Ford M, Drexler D, Fiedler T, Lentz KA, Grace JE Jr, Kolb J, Corsa J, Pierdomenico M, Jones K, Olson RE, Macor JE, Albright CF (2008) P-glycoprotein efflux and other factors limit brain amyloid beta reduction by beta-site amyloid precursor protein-cleaving enzyme 1 inhibitors in mice. J Pharmacol Exp Ther 326(2):502–513
80. May PC, Dean RA, Lowe SL, Martenyi F, Sheehan SM, Boggs LN, Monk SA, Mathes BM, Mergott DJ, Watson BM, Stout SL, Timm DE, Smith Labell E, Gonzales CR, Nakano M, Jhee SS, Yen M, Ereshefsky L, Lindstrom TD, Calligaro DO, Cocke PJ, Greg Hall D, Friedrich S, Citron M, Audia JE (2011) Robust central reduction of amyloid-beta in humans with an orally available, non-peptidic beta-secretase inhibitor. J Neurosci 31 (46):16507–16516. doi:10.1523/jneurosci.3647-11.2011
81. Hemming ML, Elias JE, Gygi SP, Selkoe DJ (2009) Identification of beta-secretase (BACE1) substrates using quantitative proteomics. PLoS One 4(12), e8477
82. Kuhn PH, Koroniak K, Hogl S, Colombo A, Zeitschel U, Willem M, Volbracht C, Schepers U, Imhof A, Hoffmeister A, Haass C, Rossner S, Brase S, Lichtenthaler SF (2012) Secretome protein enrichment identifies physiological BACE1 protease substrates in neurons. EMBO J 31(14):3157–3168
83. Wolfe MS (2011) gamma-Secretase inhibitors and modulators for Alzheimer's disease. J Neurochem 120(Suppl 1):89–98
84. Wong PC, Zheng H, Chen H, Becher MW, Sirinathsinghji DJ, Trumbauer ME, Chen HY, Price DL, Van der Ploeg LH, Sisodia SS (1997) Presenilin 1 is required for Notch1 and DII1 expression in the paraxial mesoderm. Nature 387(6630):288–292
85. Shen J, Bronson RT, Chen DF, Xia W, Selkoe DJ, Tonegawa S (1997) Skeletal and CNS defects in Presenilin-1-deficient mice. Cell 89(4):629–639
86. Searfoss GH, Jordan WH, Calligaro DO, Galbreath EJ, Schirtzinger LM, Berridge BR, Gao H, Higgins MA, May PC, Ryan TP (2003) Adipsin: a biomarker of gastrointestinal toxicity mediated by a functional gamma secretase inhibitor. J Biol Chem 278:46107–46116

87. Wong GT, Manfra D, Poulet FM, Zhang Q, Josien H, Bara T, Engstrom L, Pinzon-Ortiz M, Fine JS, Lee HJ, Zhang L, Higgins GA, Parker EM (2004) Chronic treatment with the gamma-secretase inhibitor LY-411,575 inhibits beta-amyloid peptide production and alters lymphopoiesis and intestinal cell differentiation. J Biol Chem 279(13):12876–12882
88. Doody RS, Raman R, Farlow M, Iwatsubo T, Vellas B, Joffe S, Kieburtz K, He F, Sun X, Thomas RG, Aisen PS, Siemers E, Sethuraman G, Mohs R (2013) A phase 3 trial of semagacestat for treatment of Alzheimer's disease. N Engl J Med 369(4):341–350
89. Coric V, van Dyck CH, Salloway S, Andreasen N, Brody M, Richter RW, Soininen H, Thein S, Shiovitz T, Pilcher G, Colby S, Rollin L, Dockens R, Pachai C, Portelius E, Andreasson U, Blennow K, Soares H, Albright C, Feldman HH, Berman RM (2012) Safety and tolerability of the gamma-secretase inhibitor avagacestat in a phase 2 study of mild to moderate Alzheimer disease. Arch Neurol 69(11):1430–1440
90. Golde TE, Koo EH, Felsenstein KM, Osborne BA, Miele L (2013) gamma-Secretase inhibitors and modulators. Biochim Biophys Acta 1828(12):2898–2907
91. Weggen S, Eriksen JL, Das P, Sagi SA, Wang R, Pietrzik CU, Findlay KA, Smith TE, Murphy MP, Bulter T, Kang DE, Marquez-Sterling N, Golde TE, Koo EH (2001) A subset of NSAIDs lower amyloidogenic Abeta42 independently of cyclooxygenase activity. Nature 414(6860):212–216
92. Kukar T, Prescott S, Eriksen JL, Holloway V, Murphy MP, Koo EH, Golde TE, Nicolle MM (2007) Chronic administration of R-flurbiprofen attenuates learning impairments in transgenic amyloid precursor protein mice. BMC Neurosci 8(54):54
93. Galasko DR, Graff-Radford N, May S, Hendrix S, Cottrell BA, Sagi SA, Mather G, Laughlin M, Zavitz KH, Swabb E, Golde TE, Murphy MP, Koo EH (2007) Safety, tolerability, pharmacokinetics, and Abeta levels after short-term administration of R-flurbiprofen in healthy elderly individuals. Alzheimer Dis Assoc Disord 21(4):292–299. doi:10.1097/WAD.0b013e31815d1048
94. Wang Q, Yu X, Li L, Zheng J (2014) Inhibition of amyloid-beta aggregation in Alzheimer's disease. Curr Pharm Des 20(8):1223–1243
95. Necula M, Kayed R, Milton S, Glabe CG (2007) Small molecule inhibitors of aggregation indicate that amyloid beta oligomerization and fibrillization pathways are independent and distinct. J Biol Chem 282(14):10311–10324. doi:10.1074/jbc.M608207200
96. Ballatore C, Lee VM, Trojanowski JQ (2007) Tau-mediated neurodegeneration in Alzheimer's disease and related disorders. Nat Rev Neurosci 8(9):663–672
97. Wischik CM, Edwards PC, Lai RY, Roth M, Harrington CR (1996) Selective inhibition of Alzheimer disease-like tau aggregation by phenothiazines. Proc Natl Acad Sci U S A 93 (20):11213–11218
98. Congdon EE, Wu JW, Myeku N, Figueroa YH, Herman M, Marinec PS, Gestwicki JE, Dickey CA, Yu WH, Duff KE (2012) Methylthioninium chloride (methylene blue) induces autophagy and attenuates tauopathy in vitro and in vivo. Autophagy 8(4):609–622. doi:10.4161/auto.19048
99. Medina DX, Caccamo A, Oddo S (2011) Methylene blue reduces abeta levels and rescues early cognitive deficit by increasing proteasome activity. Brain Pathol 21(2):140–149. doi:10.1111/j.1750-3639.2010.00430.x
100. Stack C, Jainuddin S, Elipenahli C, Gerges M, Starkova N, Starkov AA, Jove M, Portero-Otin M, Launay N, Pujol A, Kaidery NA, Thomas B, Tampellini D, Beal MF, Dumont M (2014) Methylene blue upregulates Nrf2/ARE genes and prevents tau-related neurotoxicity. Hum Mol Genet 23(14):3716–3732. doi:10.1093/hmg/ddu080
101. Hochgrafe K, Sydow A, Matenia D, Cadinu D, Konen S, Petrova O, Pickhardt M, Goll P, Morellini F, Mandelkow E, Mandelkow EM (2015) Preventive methylene blue treatment preserves cognition in mice expressing full-length pro-aggregant human Tau. Acta Neuropathol Commun 3:25. doi:10.1186/s40478-015-0204-4

102. Hosokawa M, Arai T, Masuda-Suzukake M, Nonaka T, Yamashita M, Akiyama H, Hasegawa M (2012) Methylene blue reduced abnormal tau accumulation in P301L tau transgenic mice. PLoS One 7(12), e52389. doi:10.1371/journal.pone.0052389
103. Wischik CM, Staff RT, Wischik DJ, Bentham P, Murray AD, Storey JM, Kook KA, Harrington CR (2015) Tau aggregation inhibitor therapy: an exploratory phase 2 study in mild or moderate Alzheimer's disease. J Alzheimers Dis 44(2):705–720. doi:10.3233/jad-142874
104. Sullivan MG (2016) Alzheimer's anti-tau drug fails phase III, but posts some benefit in monotherapy subanalysis. Clinical Neurology News, 27 July 2016
105. Fagan T (2016) In first phase 3 trial, the tau drug LMTM did not work. Period. http://www.alzforum.org/news/conference-coverage/first-phase-3-trial-tau-drug-lmtm-did-not-work-period
106. Goedert M, Spillantini MG, Cairns NJ, Crowther RA (1992) Tau proteins of Alzheimer paired helical filaments: abnormal phosphorylation of all six brain isoforms. Neuron 8 (1):159–168
107. Lee VM, Balin BJ, Otvos L Jr, Trojanowski JQ (1991) A68: a major subunit of paired helical filaments and derivatized forms of normal Tau. Science 251(4994):675–678
108. Busciglio J, Lorenzo A, Yeh J, Yankner BA (1995) beta-Amyloid fibrils induce tau phosphorylation and loss of microtubule binding. Neuron 14(4):879–888
109. Le Corre S, Klafki HW, Plesnila N, Hubinger G, Obermeier A, Sahagun H, Monse B, Seneci P, Lewis J, Eriksen J, Zehr C, Yue M, McGowan E, Dickson DW, Hutton M, Roder HM (2006) An inhibitor of tau hyperphosphorylation prevents severe motor impairments in tau transgenic mice. Proc Natl Acad Sci U S A 103(25):9673–9678
110. Trinczek B, Biernat J, Baumann K, Mandelkow EM, Mandelkow E (1995) Domains of tau protein, differential phosphorylation, and dynamic instability of microtubules. Mol Biol Cell 6(12):1887–1902
111. Hasegawa M, Smith MJ, Goedert M (1998) Tau proteins with FTDP-17 mutations have a reduced ability to promote microtubule assembly. FEBS Lett 437(3):207–210
112. Fitzgerald DP, Emerson DL, Qian Y, Anwar T, Liewehr DJ, Steinberg SM, Silberman S, Palmieri D, Steeg PS (2012) TPI-287, a new taxane family member, reduces the brain metastatic colonization of breast cancer cells. Mol Cancer Ther 11(9):1959–1967. doi:10.1158/1535-7163.mct-12-0061
113. Chesser AS, Pritchard SM, Johnson GV (2013) Tau clearance mechanisms and their possible role in the pathogenesis of Alzheimer disease. Front Neurol 4:122. doi:10.3389/fneur.2013.00122
114. Lee BH, Lee MJ, Park S, Oh DC, Elsasser S, Chen PC, Gartner C, Dimova N, Hanna J, Gygi SP, Wilson SM, King RW, Finley D (2010) Enhancement of proteasome activity by a small-molecule inhibitor of USP14. Nature 467(7312):179–184
115. Chen PC, Qin LN, Li XM, Walters BJ, Wilson JA, Mei L, Wilson SM (2009) The proteasome-associated deubiquitinating enzyme Usp14 is essential for the maintenance of synaptic ubiquitin levels and the development of neuromuscular junctions. J Neurosci 29 (35):10909–10919. doi:10.1523/jneurosci.2635-09.2009
116. Jin YN, Chen PC, Watson JA, Walters BJ, Phillips SE, Green K, Schmidt R, Wilson JA, Johnson GV, Roberson ED, Dobrunz LE, Wilson SM (2012) Usp14 deficiency increases tau phosphorylation without altering tau degradation or causing tau-dependent deficits. PLoS One 7(10), e47884. doi:10.1371/journal.pone.0047884
117. Chen HK, Liu Z, Meyer-Franke A, Brodbeck J, Miranda RD, McGuire JG, Pleiss MA, Ji ZS, Balestra ME, Walker DW, Xu Q, Jeong DE, Budamagunta MS, Voss JC, Freedman SB, Weisgraber KH, Huang Y, Mahley RW (2012) Small molecule structure correctors abolish detrimental effects of apolipoprotein E4 in cultured neurons. J Biol Chem 287(8):5253–5266. doi:10.1074/jbc.M111.276162
118. McConathy J, Sheline YI (2015) Imaging biomarkers associated with cognitive decline: a review. Biol Psychiatry 77(8):685–692. doi:10.1016/j.biopsych.2014.08.024

119. Ariza M, Kolb HC, Moechars D, Rombouts F, Andres JI (2015) Tau positron emission tomography (PET) imaging: past, present, and future. J Med Chem 58(11):4365–4382. doi:10.1021/jm5017544
120. Gaugler JE, Ascher-Svanum H, Roth DL, Fafowora T, Siderowf A, Beach TG (2013) Characteristics of patients misdiagnosed with Alzheimer's disease and their medication use: an analysis of the NACC-UDS database. BMC Geriatr 13:137. doi:10.1186/1471-2318-13-137
121. Patel KR (2015) Biogen's aducanumab raises hope that Alzheimer's can be treated at its source. Manag Care 24(6):19

Top Med Chem (2017) 24: 27–86
DOI: 10.1007/7355_2016_16

Published online: 23 December 2016

The Design, Development, and Evaluation of BACE1 Inhibitors for the Treatment of Alzheimer's Disease

Arun K. Ghosh, Emilio L. Cárdenas, and Heather L. Osswald

Abstract Alzheimer's disease (AD) is a very serious public health problem. Currently, there is no effective treatment for AD. Among the many biochemical targets for AD drug development, β-secretase (BACE1, memapsin 2) continues to be a promising drug discovery target for AD therapy. This proteolytic enzyme is a membrane-anchored aspartic acid protease that is responsible for the initial step of amyloid precursor protein (APP) cleavage, leading to the production of neurotoxic amyloid-β (Aβ) peptides in the brain. Since its identification and structural elucidation in 1999, extensive research efforts have led to the development of many promising classes of inhibitors against this protease. Structure-based design strategies led to the evolution of many small-molecule, peptidomimetic, and nonpeptide BACE1 inhibitors that have now overcome the key development challenges including selectivity and brain penetration. To date, 13 BACE1 drug candidates have been brought to clinical trials, and a number of them have advanced to phase II/III human trials. This chapter illustrates structure-based evolution of various classes of BACE inhibitors. Also, it provides a perspective on BACE1 inhibitor drugs for the treatment of AD patients.

Keywords β-Secretase, Alzheimer's disease, BACE1, BACE1 inhibitor, Drug design, Enzyme inhibitors, Memapsin 2, Nonpeptide inhibitor, Peptidomimetic inhibitor, Structure-activity relationship

A.K. Ghosh (✉)
Department of Chemistry, Purdue University, West Lafayette, IN, USA

Department of Medicinal Chemistry, Purdue University, West Lafayette, IN, USA
e-mail: akghosh@purdue.edu

E.L. Cárdenas and H.L. Osswald
Department of Chemistry, Purdue University, West Lafayette, IN, USA

Contents

Abbreviations

Aβ	Amyloid-β protein
AD	Alzheimer's Disease
ADAS	Alzheimer's Disease Assessment Scale
ADAS-Cog	Alzheimer's Disease Assessment Scale-Cognitive Subscale
ADCS	Alzheimer's Disease Cooperative Study
ADCS-ADL	Alzheimer's Disease Cooperative Study-Activities of Daily Living
ADCS-PACC	Alzheimer's Disease Cooperative Study-Preclinical Alzheimer's Cognitive Composite
ADME	Absorption distribution, metabolism, and excretion
APP	Amyloid precursor protein
BACE1	β-Site amyloid precursor protein-cleaving enzyme 1
BACE2	β-Site amyloid precursor protein-cleaving enzyme 2
BBB	Blood–brain barrier
CatD	Cathepsin D
CCS-3D	3-Domain Composite Cognition Score
CDR	Clinical Dementia Rating Scale
CDR-SB	Clinical Dementia Rating Sum of Boxes
CFI	Cognitive Function Instrument

CNS	Central nervous system
CSF	Cerebrospinal fluid
C-SSRS	Columbia-Suicide Severity Rating Scale
ECG	Electrocardiogram
FDA	Food and Drug Administration
HIV	Human immunodeficiency virus
HTS	High-throughput screening
IC_{50}	50% inhibitory concentration
K_i	Inhibitory constant
MMSE	Mini Mental State Examination
MRI	Magnetic resonance imaging
MWM	Morris water maze
NMDA	*N*-Methyl D-aspartate
NPI	Neuropsychiatric Inventory
NTB	Neuropsychological Test Battery
PET	Positron emission tomography
Pgp	P-glycoprotein
QT	Q-wave to T-wave interval
sAPPα	Soluble amyloid precursor protein α-fragment
sAPPβ	Soluble amyloid precursor protein β-fragment
SAR	Structure-activity relationship

1 Introduction

Alzheimer's disease (AD) has emerged as a serious public health issue affecting millions of elderly people around the globe. AD is an irreversible, progressive, degenerative brain disorder that slowly destroys brain function, leading to cognitive decline, behavioral change, and deterioration of bodily function in later stages. Currently, about 44 million people worldwide are believed to have Alzheimer's disease or a related dementia [1, 2]. This number is expected to grow more than 100 million cases by 2050 [3]. The global cost of current AD patient care is estimated to be a staggering $605 billion, which is equivalent to 1% of the world's gross domestic product [4]. The situation will get worse as there is no cure or disease-modifying treatment for AD at this time. A number of approved drugs are available only to treat symptoms in AD patients [5]. There is an urgent need for mechanism-based, effective AD therapies that will prevent, delay the onset, slow the progression, or improve AD symptoms.

Currently, five drugs are approved for the treatment of AD and related disorders. These include acetylcholinesterase inhibitors, tacrine (approved 1993), donepezil (1996), rivastigmine (1998), and galantamine (2001) [6, 7]. These drugs act by blocking the process that breaks down the neurotransmitter acetylcholine, a key signaling agent responsible for communication between nerve cells. In AD patients, decreased acetylcholine levels and loss of nerve cells are linked to worsening

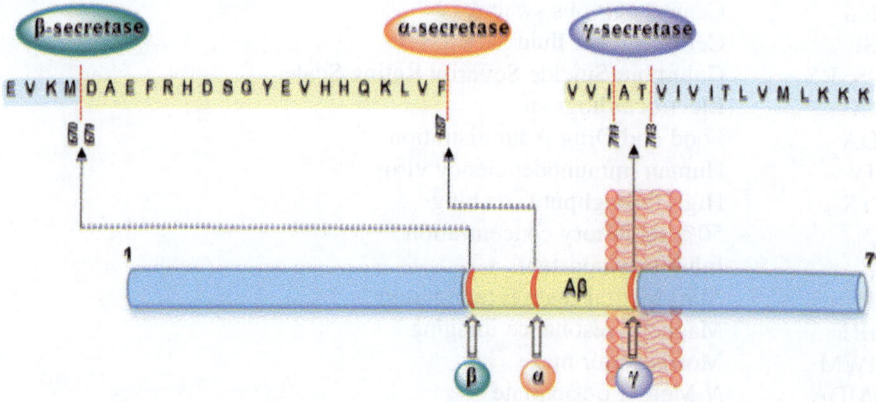

Fig. 1 Processing of APP by α-, β-, and γ-secretase. Cleavage by β-secretase provides the N-terminus of Aβ, while cleavage by γ-secretase provides the C-terminus of Aβ. Processing by γ-secretase provides both $A\beta_{40}$ and $A\beta_{42}$ by cleaving at residues 711 and 713, respectively. Cleavage by α-secretase occurs in the middle of the Aβ sequence and precludes the formation of Aβ

symptoms. Acetylcholinesterase inhibitors prevent breakdown of acetylcholine and help an maintain adequate concentration of in the brain. Another Alzheimer's drug, memantine, is an *N*-methyl D-aspartate (NMDA) receptor antagonist, approved in 2004 [8, 9]. The drug regulates the activity of glutamate in the brain. Attachment of the neurotransmitter glutamate to the cell surface of NMDA receptors allows calcium to enter the cell. This is an important event in cell signaling, as well as in learning and memory. However, in AD patients, excess glutamate released from damaged cells leads to overexposure to calcium and accelerates cell damage. Memantine disrupts this chain of events by blocking the NMDA receptors.

AD drug development has been a formidable challenge [5]. A number of current AD drug development efforts are directed towards a breakthrough AD therapy that would treat the underlying disease mechanism and halt or slow down the neuronal cell damage that causes progressive deterioration. One of the major hallmarks of Alzheimer's disease is the accumulation of neuritic plaques containing the 40–42 residue amyloid-β (Aβ) peptides and neurofibrillary tangles composed of tau protein in the brain [1, 10]. Since a high brain Aβ level is regarded as an important factor in AD pathogenesis, clinical intervention to reduce its production has become a logical approach for AD therapy development [11, 12]. The origin of neuritic plaques can be attributed to proteolytic processing of amyloid precursor protein (APP) (Fig. 1). More specifically, the sequential cleavage of APP by β-secretase (BACE1, memapsin 2) and γ-secretase yields Aβ [13, 14]. The two main Aβ products of APP processing, $A\beta_{40}$ and $A\beta_{42}$, both play a key role in the aggregation of neuritic plaques. While $A\beta_{42}$ is more prone to aggregation, increasing levels of both peptides are observed early in the pathology of AD, and these rising levels are correlated with the severity of dementia observed in AD patients [15].

The preceding plaques are shown to be neurotoxic and eventually lead to neuronal death [16]. Therapeutic inhibition of β- or γ-secretase has been pursued in AD drug development since these two proteases are the first and second steps in the APP processing leading to Aβ production [11, 17]. Following the cloning and identification of BACE1 in 1999, significant research efforts led to the development of several classes of potent and selective BACE1 inhibitors as a potential treatment of Alzheimer's disease [18, 19]. Herein, we outline structure-based evolution of various classes of BACE1 inhibitors, clinical evaluation of these inhibitors, and their perspective in the treatment of AD.

1.1 *The Amyloid Hypothesis of AD*

Most AD patients have cases of sporadic AD; however, genetic predispositions to AD can occur within families. While both sporadic and familial AD are pathologically the same, familial AD tends to occur as early-onset AD, with the symptoms appearing as early as age 30 [1]. Familial AD is caused by mutations in either APP or the presenilins responsible for γ-secretase activity. APP mutations result in more efficient cleavage by BACE1, increased propensity for aggregation, or an increase in the more aggregation-prone $A\beta_{42}$ [20]. There are many more presenilin mutations known, which lead to increased γ-secretase processing of APP into pathogenic $A\beta_{42}$ over the more phenotypically normal $A\beta_{40}$. These mutations in early-onset AD and similar pathology of sporadic AD have led to the amyloid hypothesis of AD progression (Fig. 2).

The amyloid hypothesis of AD proposes that the formation of Aβ plaques initiates the flow of outcomes that ultimately leads to dementia in AD patients

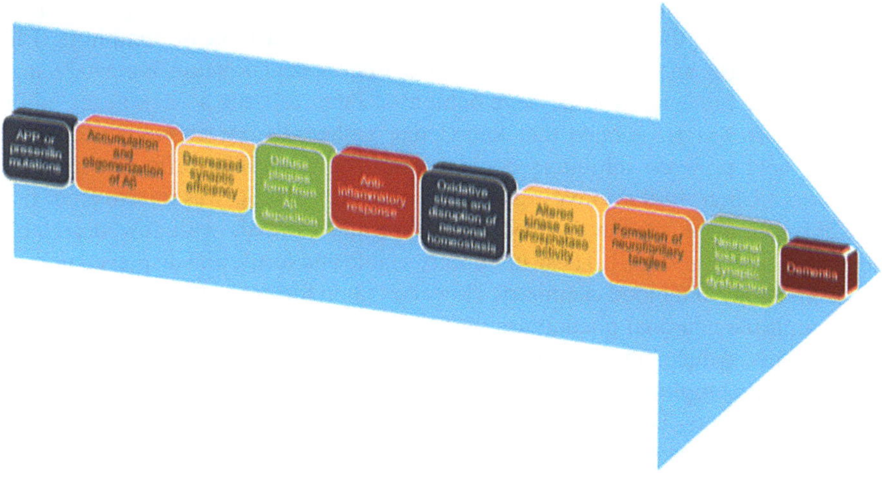

Fig. 2 Progression of events postulated by the amyloid cascade hypothesis

[21]. The accumulation of Aβ peptides leads to their oligomerization. This event results in decreased activity at the neuronal synapses. These soluble oligomers further associate to form Aβ plaques. These plaques trigger an anti-inflammatory response causing oxidative stress of the neurons. This can disrupt normal kinase and phosphatase activity, resulting in the hyperphosphorylation of tau protein and subsequent neurofibrillary tangle formation. This adds to abnormal signaling and further synaptic impairment. Ultimately, these events result in neuronal death and dementia in the AD patient [22].

While the occurrence of familial AD and the genetic mutations associated with familial AD are strong arguments for the amyloid hypothesis, some observations remain unexplained [23]. It was once suggested by histological studies that the number of Aβ plaques in a diseased brain does not directly correlate to the degree of cognitive degeneration in AD patients. However, biochemical and later histological studies have shown that there is a correlation between plaque formation and cognitive decline [15, 23]. While there are further questions regarding the exact neuronal toxicity of Aβ in vivo, the amyloid hypothesis is still broadly accepted as the general pathological cascade of events in AD [21].

2 BACE1 as a Target for AD

BACE1 is an aspartic acid protease that plays a prominent role in the generation of neurotoxic Aβ plaques which has been strongly associated with AD pathology (Fig. 3) [24–27]. In light of the initial cloning and further elucidation of the role of BACE1 in AD pathology [26, 28], it has been determined that the inhibition of BACE1 may serve as a promising therapy for AD.

The proteolytic cleavage of APP by BACE1 marks the initial step of amyloidogenesis, which in turn triggers the later events of AD pathogenesis [29–31]. Additionally, the development of successful clinical protease inhibitors (PIs) for HIV protease, another extensively studied aspartic acid protease, has established an invaluable precedent for the development of other aspartic acid protease inhibitors [32]. It was envisioned that knowledge gathered from previous successful aspartic acid protease inhibitors could be utilized to develop effective BACE1 inhibitors. Upon further investigation of the enzyme, it was discovered that BACE1-null mice appeared to be phenotypically normal, with few detectable abnormalities [30, 31, 33]. These mice were viable and fertile and displayed normal tissue histology and blood chemistry. However, upon further analysis of BACE1$^{-/-}$ mice, multiple neuronal phenotypes were discovered, including seizures, memory impairments, reduced myelination, and schizophrenia, among others [34–39]. It is possible that complete cessation of BACE1 activity may lead to unwanted mechanistic side effects. The establishment of a proper therapeutic window would allow for proper inhibition of BACE1 in order to slow or halt the progression of AD while allowing other necessary BACE1 functions to continue.

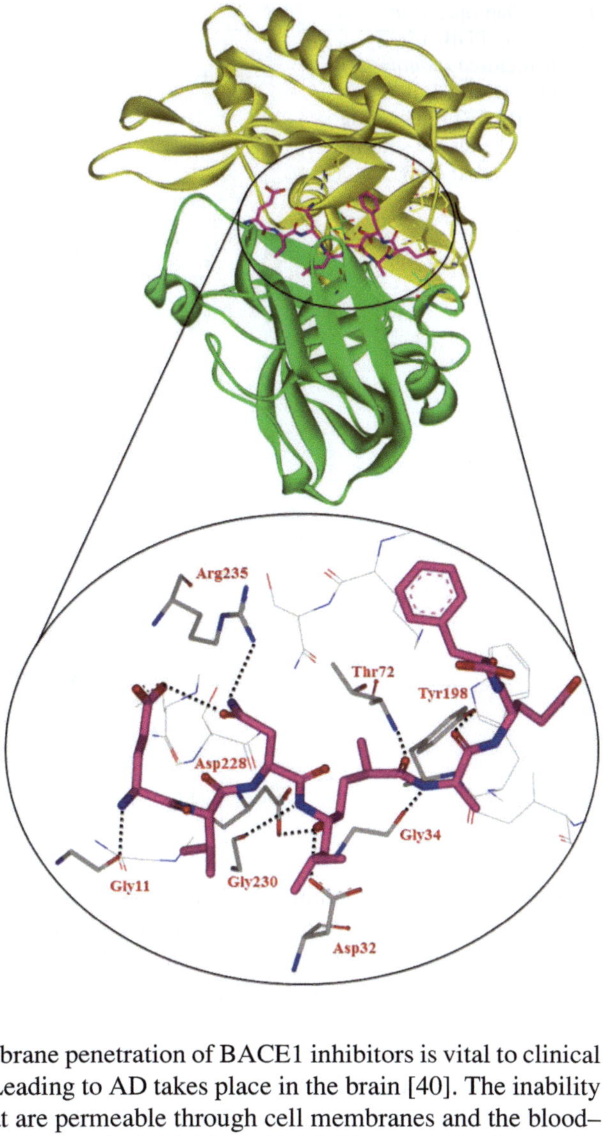

Fig. 3 Overall structure of BACE1 and interactions of inhibitor OM99-2 (**2**) in the BACE1 active site (PDB: 1FKN)

The propensity for membrane penetration of BACE1 inhibitors is vital to clinical success as Aβ production Leading to AD takes place in the brain [40]. The inability to generate compounds that are permeable through cell membranes and the blood–brain barrier (BBB) would discredit the inhibition of BACE1 as a viable therapy for AD. Validation studies have shown that peptide and sterol linkers have allowed inhibitors to penetrate sites of Aβ accumulation as well as diminish Aβ brain levels in AD mice [41, 42]. However, once inhibitors enter into the brain, they face ejection via P-glycoprotein (Pgp). Knockout studies have shown that Pgp-null mice treated with brain-penetrant BACE1 inhibitors resulted in a decline in Aβ levels, while wild-type mice did not benefit from such treatment [43].

The structural features of BACE1 also provide a challenge when it comes to inhibitor design. The type 1 transmembrane protein is characterized by its

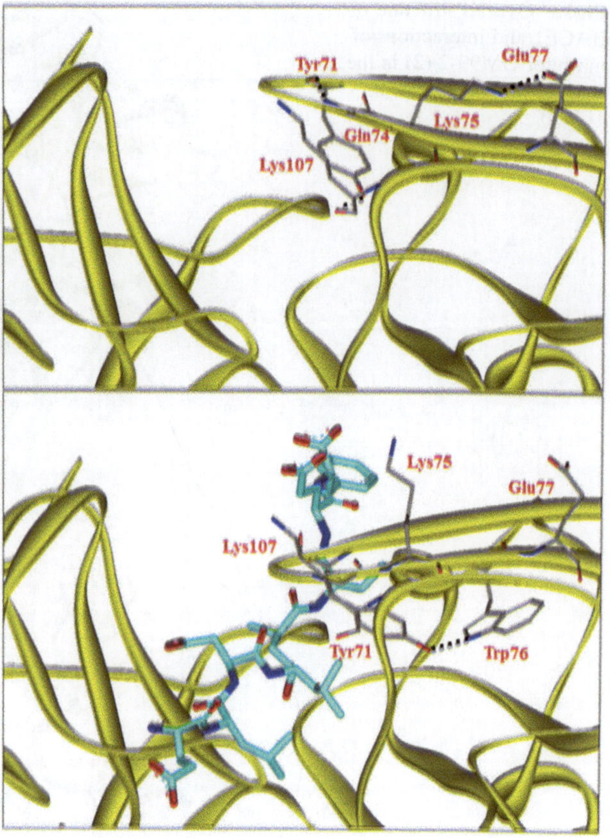

Fig. 4 Flap-open (*top*, apoprotein, PDB: 1SGZ) and flap-closed (*bottom*, bound to OM00-3, PDB: 1M4H) BACE1. Stabilizing hydrogen bonds are shown as *black dashed lines*

expansive catalytic domain which is marked by the centrally located catalytic aspartates Asp32 and Asp228. Free BACE1 features a flap-open conformation that is energetically stable due to the multiple hydrogen bonds in the flap region of the enzyme (Fig. 4, top) [44]. When a substrate is bound, BACE1 assumes a flap-closed conformation, which results in the loss of these optimal hydrogen bonds (Tyr71-Gly74, Lys75-Glu77, Tyr71-Lys107) (Fig. 4, bottom). Interaction with the substrate, along with the formation of a new hydrogen bond between Tyr71 and Trp76, helps to stabilize this conformational shift [44]. In the open conformation, substrates are allowed to enter the active site of the enzyme via a cleft. However, the presence of a bottleneck within the enzyme inherently favors flexible substrates. It is thought that this bottleneck feature may serve as a factor in substrate selectivity [44]. Meanwhile, the somewhat broad specificity in most of the subsites along the active site poses a major hurdle in inhibitor design efforts. X-ray crystal structures of BACE1 validate that the active site is part of a long cleft that serves as a template for potential interactions with designed inhibitors [45, 46].

3 Evolution of BACE1 Inhibitors

BACE1 is an aspartic acid protease which possesses two highly conserved aspartic acid residues in the active site for catalytic cleavage of its peptide substrate, APP. Its catalytic mechanism involves the coordination of a water molecule between the active site aspartates, where one aspartate activates the water molecule for nucleophilic attack on the carbonyl carbon of the substrate scissile site. The resulting tetrahedral intermediate rearranges to provide two peptide fragments. All current aspartic acid protease inhibitor drugs are designed based upon a transition-state mimetic concept which incorporates hydroxyethylene isosteres at the scissile site [19]. Since the catalytic aspartates of BACE1 are similar to other aspartic acid proteases, such as HIV protease and renin, similar principles were utilized for BACE1 inhibitor design at the outset [19]. BACE1 cleaves APP and generates Aβ mainly in the endosomes of neurons. Therefore, for BACE1 inhibitors to be clinically effective, they must have the necessary properties to cross the BBB and neuronal membranes. Furthermore, clinically relevant inhibitors must exhibit a promising drug-like absorption, distribution, metabolism, and excretion (ADME) profile. Along with good ADME properties, inhibitors must establish selectivity against other aspartic acid proteases. BACE2 is the closest homologue of BACE1 with 64% of the primary structure conserved within both proteases [47]. Cathepsin D (CatD) is a significant off-site target, as it is the most abundant aspartic acid protease found in human cells [48]. Despite many challenges in the design of effective BACE1 inhibitors, there has been enormous progression on this front. Many classes of preclinical inhibitors with impressive potency and selectivity have evolved over the years. As many as 13 drug candidates have entered into clinical development. Herein, we provide a brief overview of the structural evolution of drug-like BACE1 inhibitors.

3.1 Structure-Based Design of Inhibitors

The catalytic properties of BACE1 were thoroughly investigated following its cloning. It has an elongated substrate-binding site that can accommodate up to 11 substrate residues. BACE1 shows a broad specificity. The initial specificity and knowledge of kinetics were obtained from the hydrolysis of several peptides [26, 49]. It is noteworthy that Swedish APP (SEVNL/DAEFR) gets cleaved by BACE1 40-times faster than wild-type APP (SEVKM/DAEFR). Furthermore, specificity studies revealed that alanine is preferred at P1′. This information was used for the design of the first potent substrate-based inhibitors containing seven and eight residues with a nonhydrolyzable Leu-Ala-based hydroxyethylene dipeptide isostere [50]. Inhibitor **1** ($K_i = 36$ nM) is considerably less potent than inhibitor **2** ($K_i = 1.6$ nM), which contains eight residues (Fig. 5). A statin-substituted

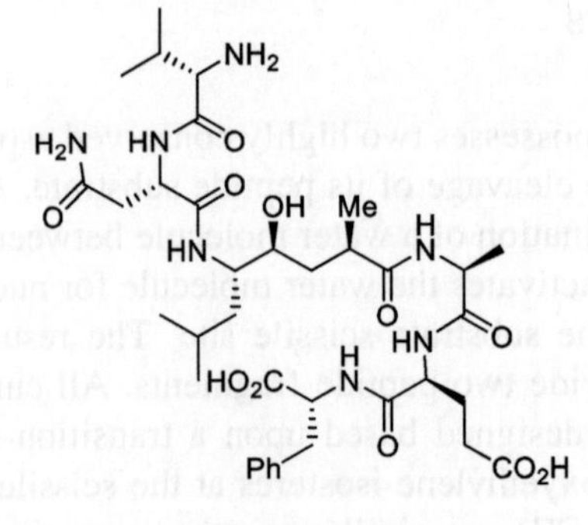

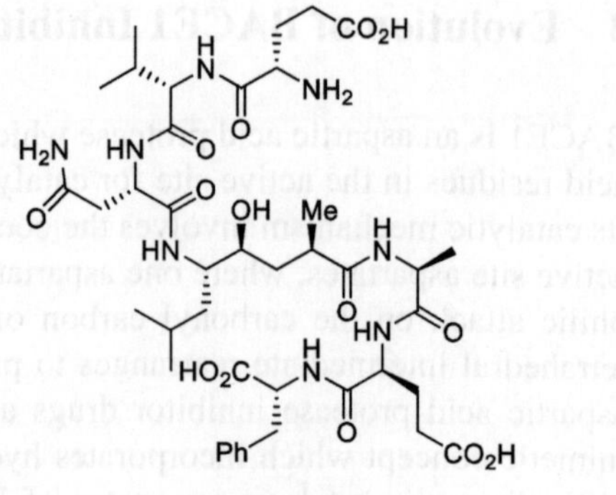

1 (OM99-1)
K_i = 36 nM (MW 764 Da)

2 (OM99-2)
K_i = 1.6 nM (MW 893 Da)

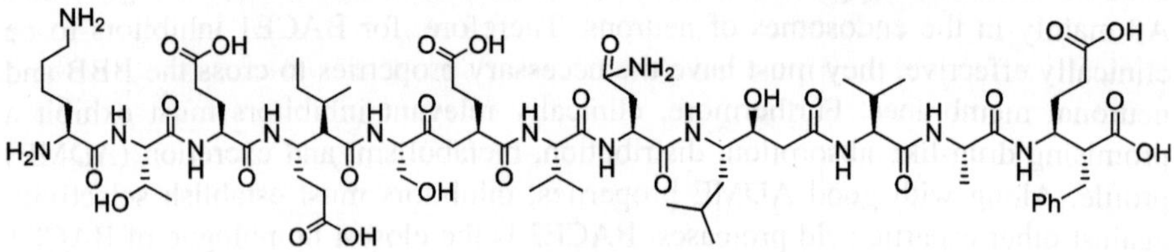

3 (StatVal)
K_i = 30 nM (MW 1564 Da)

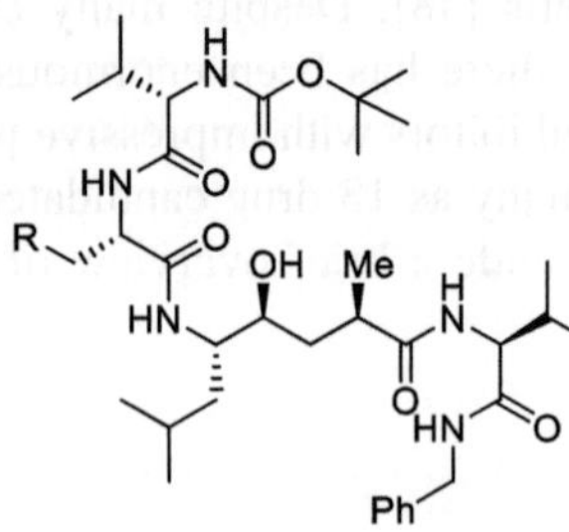

4 R = $CONH_2$
K_i = 5.9 nM (MW 704 Da)

5 R = CH_2SMe
K_i = 2.5 nM (MW 722 Da)

Fig. 5 Structures of pseudopeptidic inhibitors **1–5**

inhibitor was designed by incorporating APP residues. Inhibitor **3** ($IC_{50} = 30$ nM) shows similar inhibitory activity as inhibitor **1** [25].

The X-ray structure of the protease domain of human BACE1 bound to inhibitor **2** was determined at 1.9 Å resolution [45]. This X-ray structure provided critical molecular insight into the ligand-binding site interactions in the active site. The structure revealed that six of the eight residues (P2′–P4) are bound in the active site in an extended confirmation. The catalytic aspartates Asp32 and Asp228 form four hydrogen bonds with the hydroxyl group of the transition-state isostere. The inhibitor also forms ten additional hydrogen bonds throughout the active site including the flap region of BACE1.

The structure shows a rare kink in the S2′ subsite where the P2′ carbonyl forms a hydrogen bond with the Tyr198 hydroxyl group. Therefore, this interaction may be exploited for selectivity design. Interestingly, both the P3′- and P4′-side-chain residues do not form any specific interactions with BACE1, suggesting that these ligands can be further optimized. Based upon this molecular insight, inhibitor **4**, with an *N*-benzyl amide in place of P3′ and P4′ residues, showed potent BACE1 activity [51]. The inhibitor makes a number of critical hydrogen bonding interactions in the S2–S4 subsites. These include a hydrogen bond between P2-Asn with Arg235 in the S2 subsite and an intramolecular hydrogen bond with P4-Glu and P2-Asn, which may facilitate hydrogen bonding with Arg235. Structure-based replacement of the P2 ligand resulted in a number of potent inhibitors with reduced molecular weight and peptidic features. Inhibitor **5** is nearly as potent as inhibitor **2** [51].

3.2 Design of Peptidomimetic Inhibitors

The design of substrate-based inhibitors and their X-ray structural studies enabled structure-based design of numerous highly potent peptidomimetic inhibitors with reduced molecular weight and drug-like properties. A variety of dipeptide isosteres, including hydroxyethylene, hydroxyethylamine, and statins, were integrated into the design of these inhibitors. A number of previous reviews cover the evolution of these inhibitors in detail [11, 32]. Herein, we provide a brief highlight of peptidomimetic inhibitors that have addressed specific issues such as reduction of peptidic characteristics, improved cellular and in vivo properties, and design of selective inhibition which are critical to drug development.

3.2.1 Hydroxyethylene-Based Inhibitors

Extensive research efforts in the development of peptidomimetic BACE1 inhibitors led to compounds with molecular weights of 550–650 Da. A number of BACE1 inhibitors have shown good selectivity against other human aspartic acid proteases, particularly against two of the eight aspartic acid proteases deemed necessary, BACE2 and CatD. BACE2 selectivity may be critical, as it has similar specificity as BACE1. The other aspartic acid protease, cathepsin D, is highly abundant in cells, and therefore lack of selectivity will greatly reduce drug concentration of BACE1 inhibitors.

Structure-based design of highly selective inhibitors is shown in Fig. 6. Compound **6** exhibited enhanced selectivity while maintaining the potency of previous inhibitors ($K_i = 0.3$ nM) [52]. It displayed an impressive 436-fold selectivity over CatD and 1,186-fold selectivity over BACE2. The elucidation of a **6**-bound X-ray

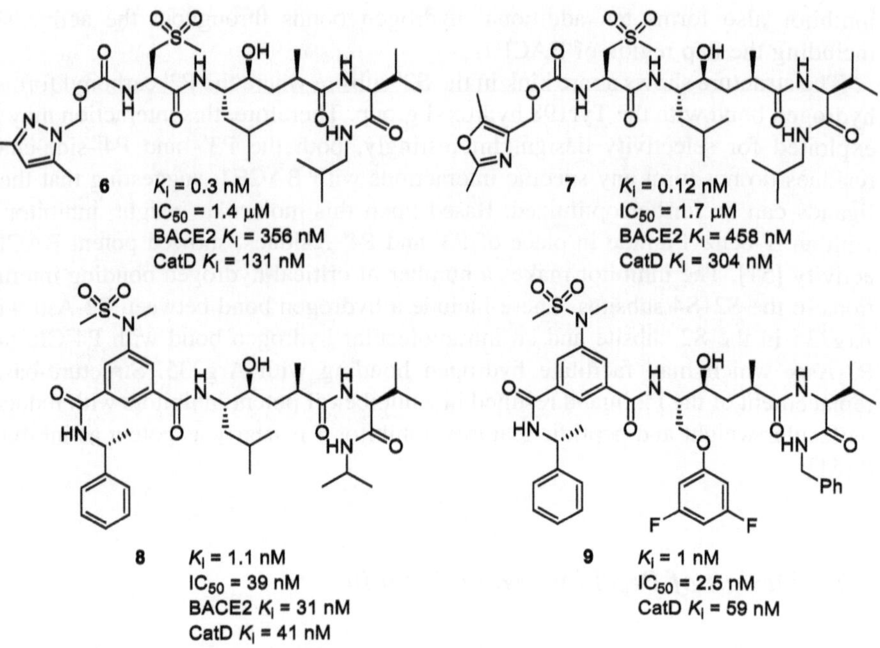

Fig. 6 Structures and activity of hydroxyethylene-based inhibitors **6–9**

crystal structure of BACE1 provided important insight of this observed selectivity. As shown in Fig. 7, the key interactions were observed between Thr232 and a nitrogen within the pyrazole heterocycle, while methyl substituents located on the heterocycle were found to also make interactions within the hydrophobic S3 subsite. The P2 sulfone made other significant hydrogen bonding interactions with Arg235 and a water-mediated hydrogen bond to Thr232 [52]. Substitution of the pyrazolylmethyl ring with the corresponding oxazolylmethyl ligand furnished compound **7**, which was found to have improved potency ($K_i = 0.12$ nM) as well as significantly improved selectivity against BACE1 (3,800-fold over BACE2 and 2,500-fold over CatD) [52]. Though inhibitors **6** and **7** were able to display enhanced selectivity, they exhibited IC_{50} values in the micromolar range (1.4 μM and 1.7 μM, respectively) in Chinese hamster ovary cells [52]. Since Aβ IC_{50} and BBB penetration are important features for viable AD therapies, further design modifications must be made to improve subcellular membrane permeability.

In an effort to improve cellular inhibitory activity, inhibitor **8** was designed by incorporating isophthalamide derivatives within the S2 subsite [53]. Inhibitor **8** displayed a K_i of 1.1 nM and a cellular IC_{50} of 39 nM in Chinese hamster ovary cells. Molecular modeling showed that the *N*-methylsulfonamide substituent on the P2 isophthalamide fit nicely in the S2 site. This ligand can make extensive hydrogen bonding interactions with Asn233, Ser325, and Arg235. The model also revealed that the (*R*)-α-methylbenzylamide moiety at the P3 site makes favorable hydrophobic contacts in the S3 pocket [53]. Intraperitoneal injections of 8 mg/kg of

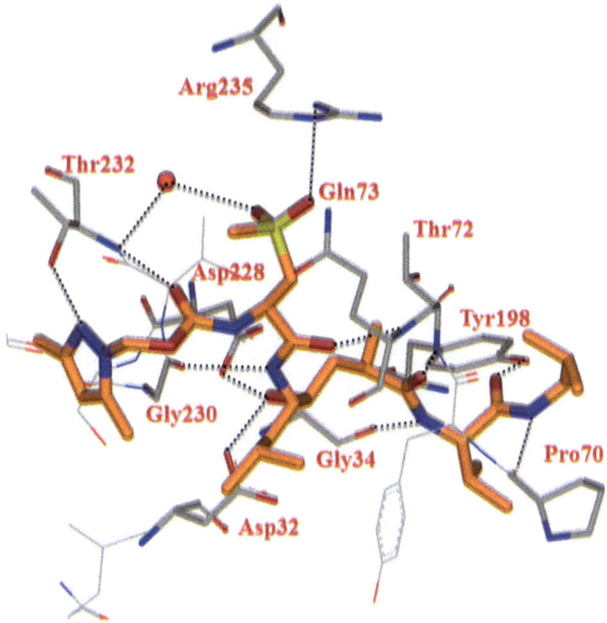

Fig. 7 X-ray crystal structure of **6**-bound BACE1 (PDB: 2G94)

compound **8** reduced plasma Aβ_{40} in Tg2576 transgenic mice by 30% after 8 h. Inhibitor **8** also displayed a 28-fold selectivity over BACE2 and 37-fold selectivity over CatD [53]. Inhibitor **9** featuring a difluorophenyl ether moiety as the P1 ligand proved to be potent with an enzymatic IC_{50} of 2.5 nM [54].

3.2.2 Hydroxyethylamine-Based Inhibitors

Hydroxyethylamine transition-state mimics have been widely utilized in the design of several FDA-approved HIV-1 protease inhibitors [55, 56]. BACE1 inhibitors with hydroxyethylamine isosteres have been investigated extensively. An early example of a pseudopeptide-based inhibitor containing a hydroxyethylamine isostere is compound **10** in Fig. 8 [57]. This pseudopeptidic inhibitor incorporated a P1 leucine side chain and a transition-state hydroxyl group with an (*S*)-configuration. Further structure-activity relationship studies revealed that a P1-phenylalanine side chain and a (*R*)-hydroxyl configuration are preferred by the BACE1 active site. Inhibitor **11**, with a phenylalanine P1 side chain and a (*R*)-hydroxyl configuration, showed good BACE1 inhibitory activity as well as Aβ cellular inhibitory activity [58]. It exhibited a 15-fold selectivity over BACE2, 500-fold selectivity over CatD, and greater than 3,000-fold selectivity over renin.

Inhibitor **12** incorporated a lipophilic 3-methoxybenzyl group as the P1 ligand and a phenylalanine side chain as the P1′ ligand. This inhibitor exhibited excellent properties, showing an enzymatic K_i of 1.8 nM and cellular IC_{50} of 1.0 nM [59, 60]. An X-ray crystal structure of **12**-bound BACE1 was determined at 2.05 Å resolution.

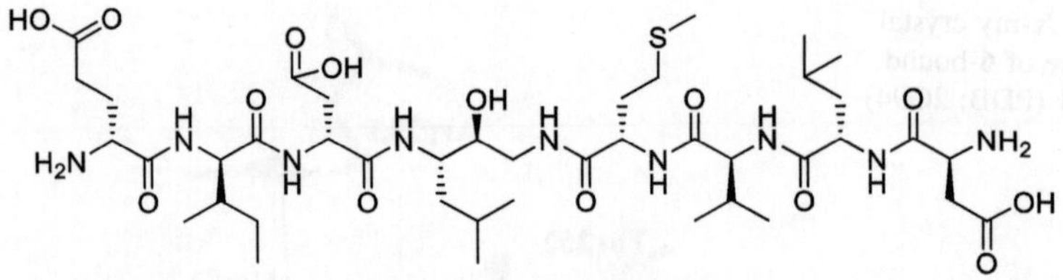

10 IC_{50} = 49 nM

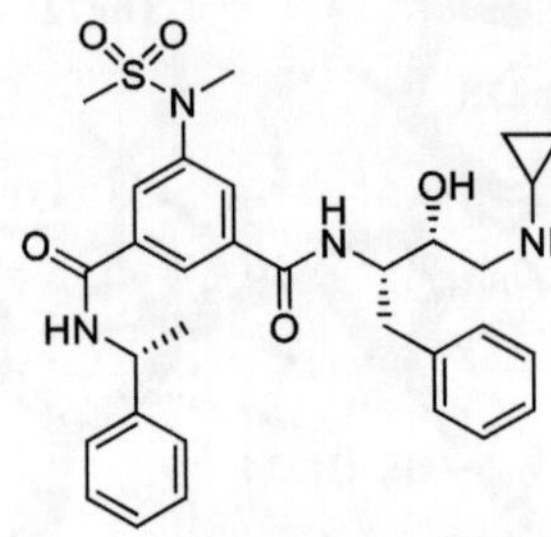

11 IC_{50} = 15 nM
cell IC_{50} = 29 nM
BACE2 IC_{50} = 230 nM
CatD IC_{50} = 7620 nM

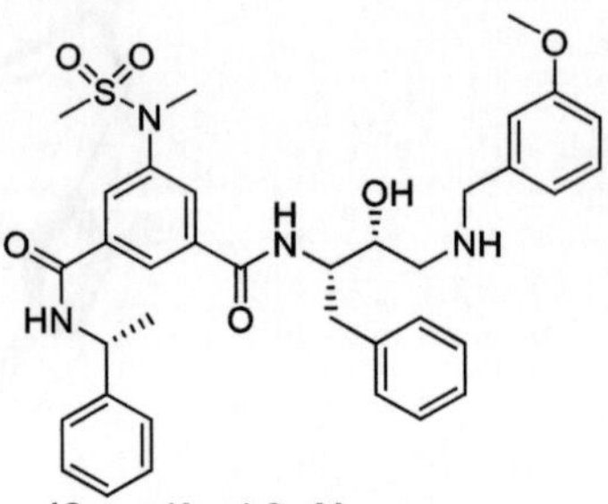

12 K_i = 1.8 nM
IC_{50} = 1 nM
BACE2 K_i = 137 nM
CatD K_i = 81 nM

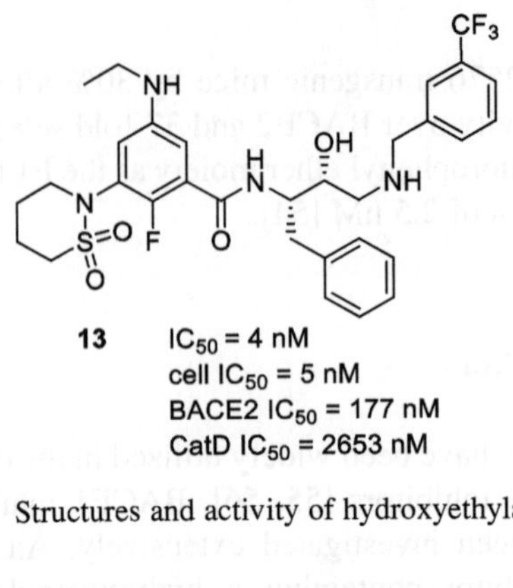

13 IC_{50} = 4 nM
cell IC_{50} = 5 nM
BACE2 IC_{50} = 177 nM
CatD IC_{50} = 2653 nM

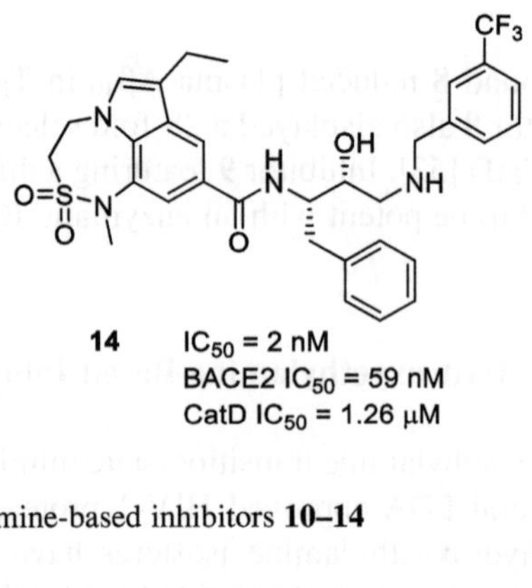

14 IC_{50} = 2 nM
BACE2 IC_{50} = 59 nM
CatD IC_{50} = 1.26 μM

Fig. 8 Structures and activity of hydroxyethylamine-based inhibitors **10–14**

As shown in Fig. 9, incorporation of the P1′ methoxybenzylamine resulted in extensive hydrophobic interactions with Gly34, Ile126, Tyr198, Pro70, Tyr71, and Thr72. The transition-state hydroxyl group formed two hydrogen bonds with Asp32; the other catalytic aspartic acid, Asp228, was bound by the hydroxyethylamine nitrogen. Both oxygens of the sulfonamide moiety of the P2 isophthalamide group formed a network of hydrogen bonds with Ser325, Asn233, and Thr232 [60].

Inhibitor **12** showed a 23-fold selectivity over CatD and 39-fold selectivity over BACE2. An intraperitoneal administration of this inhibitor with a dose of 8 mg/kg in Tg2576 mice resulted in a 65% decline of $A\beta_{40}$ levels [60]. An intravenous administration of 4 μg/g ^{14}C-radiolabeled **12** in rats showed the presence of 16% of the inhibitor in the brain over 24 h, demonstrating the BBB

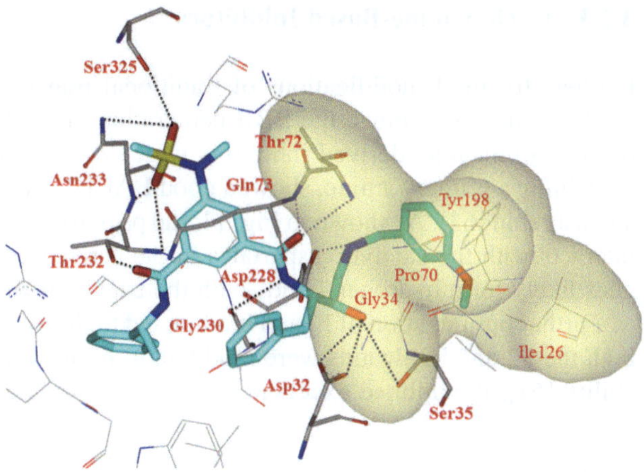

Fig. 9 X-ray crystal structure of **12** in the BACE1 active site (PDB: 2VKM)

penetrability of inhibitor **12** [59]. Further in vivo cognitive studies were carried out by installation of an osmotic pump in transgenic Tg2576 mice. The inhibitor was administered at 33.4 μg/g/day. Four experiment groups at varying ages and treatment for varying lengths of time were investigated. The first cohort began treatment with 5.5-months old and sustained treatment for 6.7 months. Cognitive performance, as measured by the Morris water maze (MWM), did not improve at 1.5 or 4.6 months. However, at 6.7 months, the treated group performed significantly better than the control [59]. In older mice (aged 16 months), the administration of inhibitor **12** did not provide any significant cognitive rescue after 4.4 months of treatment, despite the reduction of Aβ levels similar to that of younger cohorts [59]. These results indicated that even a partial reduction of Aβ production and neuritic plaque was sufficient to rescue cognitive decline in younger transgenic AD mice. There were no observable signs of toxicity or evidence for the accumulation of unprocessed APP [59].

Inhibitor **13**, with a six-membered sultam, showed good BACE1 IC_{50} and cellular Aβ activity [61]. The *meta*-ethylamine located in the S3 site provided optimum selectivity and potency with a 44-fold selectivity over BACE2 and a 663-fold selectivity over CatD. A fluorine at the C2 position of the isophthalic aryl ring improved potency in cells expressing both wild-type and Swedish-mutant APP ($IC_{50} = 5$ nM and 40 nM, respectively) [61]. Inhibitor **13** showed modest oral bioavailability in transgenic mice studies. The administration of oral doses of 250 mg/kg twice daily showed as much as a 23% reduction in $A\beta_{42}$ levels within the diseased mouse brain. When dosed alongside a Pgp inhibitor, $A\beta_{42}$ reduction increased to 55% [61]. Tricyclic sultam derivatives, as represented by inhibitor **14**, were developed in order to optimize the metabolic stability of inhibitors. This inhibitor showed a BACE1 IC_{50} value of 2 nM and modest selectivity over BACE2 and CatD [62].

3.2.3 Carbinamine-Based Inhibitors

Further structural modifications of traditional transition-state isosteres were pursued in order to optimize the BBB penetration of inhibitors. As shown in Fig. 10, potent carbinamine-derived inhibitors were designed and evaluated. Inhibitor **15**, containing a primary amine, showed good BACE1 activity and excellent selectivity against renin [63]. As shown in Fig. 11, the primary amine functionality presumably interacts with the catalytic aspartates in the active site. The two oxadiazole nitrogens likely form hydrogen bonds with the backbone of Gln73 and Thr72.

Further optimization of ligand binding led to the evolution of inhibitor **16**, where both the P2 and P3 ligands were modified to improve potency and cellular permeability [64]. It exhibited excellent potency, yet its selectivity against BACE2 was

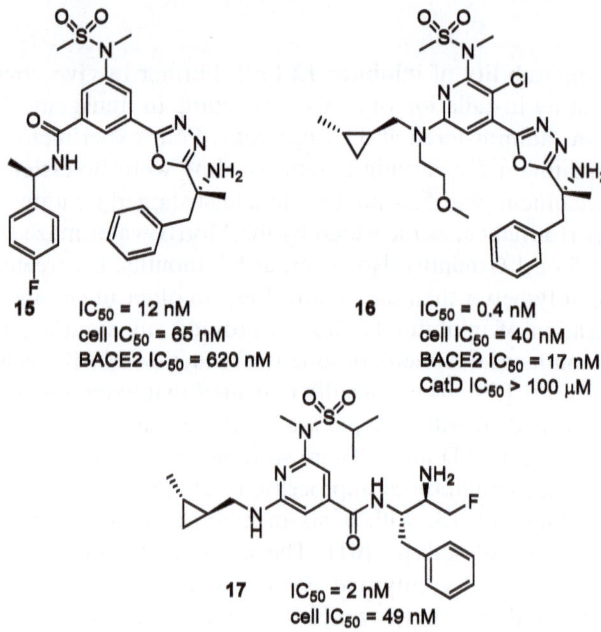

Fig. 10 Structures and activity of carbinamine-based inhibitors **15–17**

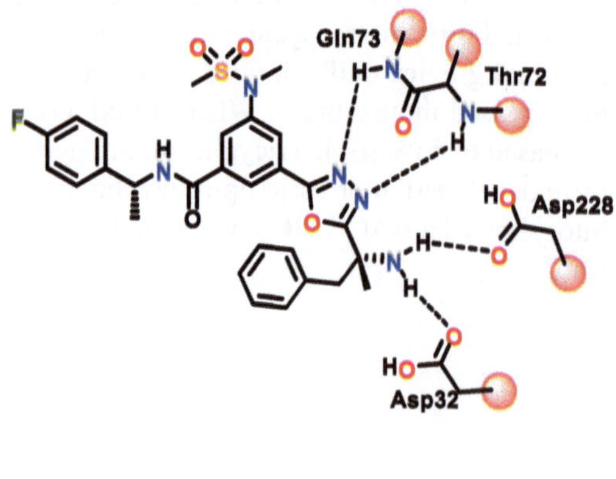

Fig. 11 Binding mode of BACE1 inhibitor **15**

moderate (42-fold). However, it showed enhanced selectivity against CatD. Intraperitoneal administration of **16** at a 100 mg/kg dose resulted in a 26% decline of postinjection $A\beta_{40}$ levels. Further, time- and dose-dependent studies in rhesus monkeys showed a 65% reduction of plasma $A\beta_{40}$ levels after 4 h, a return of levels after 8 h, and a complete recovery of $A\beta_{40}$ after a 24-hour period. Inhibitor plasma concentration was determined to be 3.8 μM. The oral bioavailability of the inhibitor was found to be poor. However, when dosed with pharmacokinetic booster ritonavir, the bioavailability of the inhibitor improved to 83% [64, 65].

Inhibitor **17** incorporated the 2,6-diaminoisonicotinamide core of inhibitor **16**. It showed potent BACE1 activity and cellular Aβ inhibitory activity of 49 nM [66]. In vivo studies in transgenic mice using 50 mg/kg i.v. bolus showed decreased $A\beta_{40}$ levels in the brain by 34% at 3 h post-dose [66]. Pharmacokinetic evaluation in rats showed a high clearance and volume of distribution (Cl = 45.8 mg/min/kg and Vd = 3.9 l/kg). Unfortunately, the half-life of inhibitor **17** in vivo was very short, and the bioavailability was poor ($t_{1/2} = 1.6$ h and $\%F = 13\%$) [66].

3.2.4 Reduced Amide-Based Inhibitors

X-ray structure-based design of BACE1 inhibitors with a reduced amide isostere provided very potent and selective inhibitors. As shown in Fig. 12, inhibitor **18** showed cellular Aβ inhibitory activity of 22 nM [67]. Subsequent optimization of ligand binding, particularly in the S1′ subsite, led to very potent and highly selective BACE1 inhibitors. As shown in inhibitor **19**, an (*S*)-hydroxyl group was incorporated to form hydrogen bonds with substituents in the S1′ subsite [68]. It exhibited a 7,000-fold selectivity against BACE2 and 250,000-fold selectivity against CatD. The corresponding inhibitor with a (*R*)-configuration on the P1′ hydroxyl group was significantly less potent and selective. The X-ray crystal structure of **19**-bound BACE1 was determined at 2.2 Å resolution. As shown in Fig. 13, the reduced amide isostere formed two tight hydrogen bonds with the active site aspartates. The allo-threonine hydroxyl group is within proximity to form a hydrogen bond with the Tyr198 side chain hydroxyl group. Since this interaction is absent in inhibitor **18**, the enhanced potency and selectivity of inhibitor **19** may be due to specific interactions with Tyr198. Further structural modification of the P2 ligand with a 7,6,5-tricyclic indole resulted in very potent and selective inhibitors as represented in inhibitor **20**. No in vivo studies were reported for these inhibitors.

3.2.5 Macrocyclic Peptidomimetic Inhibitors

The large active site of BACE1 provides an opportunity to design macrocyclic BACE1 inhibitors with a prearranged bioactive conformation. As shown in the X-ray structure of substrate-based inhibitor **2**-bound BACE1 (Fig. 1), the S1–S3 subsites are open in nature. Thus, macrocyclic rings formed between the P1 ligand

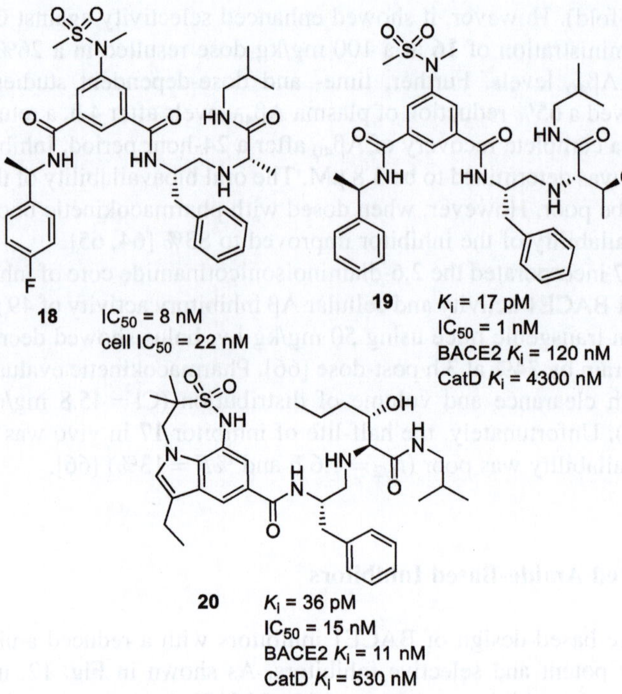

Fig. 12 Structures and activity of reduced amide-based inhibitors **18–20**

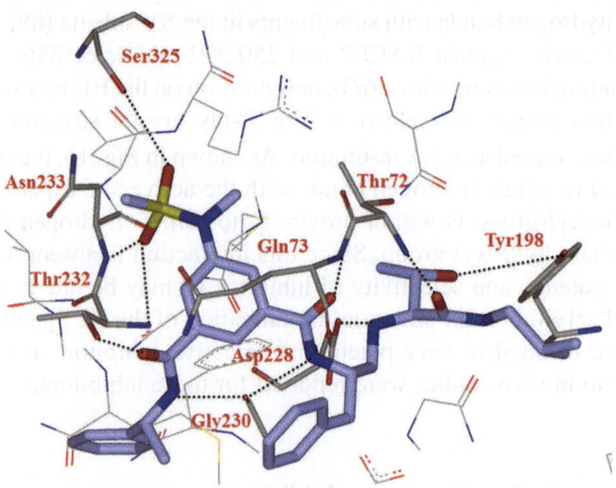

Fig. 13 X-ray crystal structure of inhibitor **19**-bound BACE1 (PDB: 4GID)

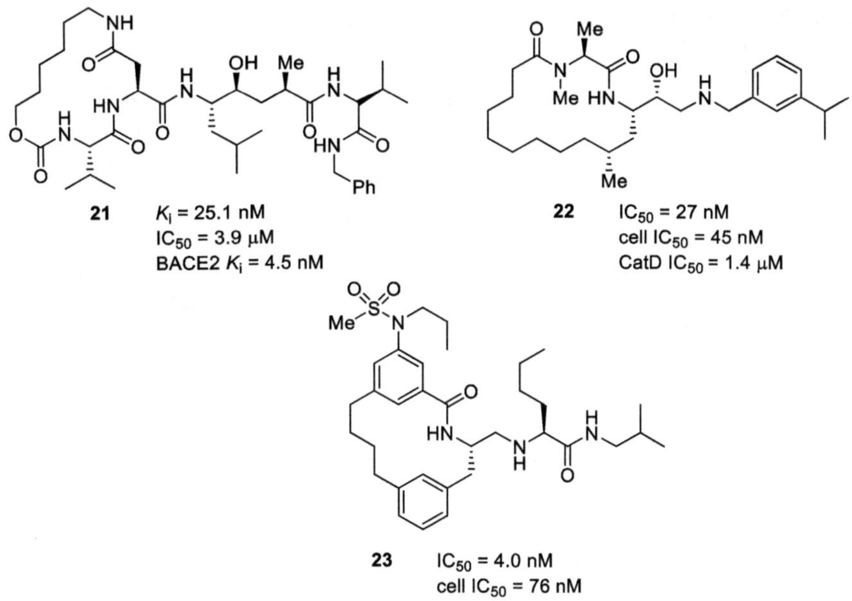

Fig. 14 Structures and activity of peptidomimetic macrocyclic-based inhibitors **21–23**

and the nitrogen of a P2 amide can be accommodated. Macrocyclic inhibitors may lead to more selective inhibitors as the bond rotations are restricted.

Initial examples took advantage of the X-ray structural insight provided by the substrate-based inhibitor **2**-bound BACE1, as shown in Fig. 14. Inhibitor **21**, with a 16-membered macrocycle containing an amide and urethane functionality, showed good BACE1 enzymatic activity [69]. Cyclic inhibitors were more potent than acyclic inhibitors. Also, studies showed that ring sizes 14–16 can be accommodated. An X-ray structure of inhibitor **21**-bound BACE1 showed key ligand-binding site interactions. As shown in Fig. 15, the Leu-Ala dipeptide isostere binds similarly to inhibitor **2**. Furthermore, the P2 asparagine on the macrocycle is able to interact with Arg235 in the S2 subpocket. This class of inhibitors, however, did not show improved cellular activity or selectivity against BACE2.

Inhibitor **22,** with a hydroxyethylamine isostere, showed improved cellular permeability [70]. There was also a modest improvement in selectivity, showing a 63-fold selectivity over CatD. Macrocyclic inhibitors with a reduced amide isostere were designed to bolster the pharmacokinetic profile. The 14-membered macrocyclic inhibitor **23** showed good BACE1 activity and cellular Aβ inhibitory potency [71]. The inhibitor exhibited a much improved Pgp ratio of 5.5. Further, in vivo studies showed that a 100 mg/kg dose of **23** cleared brain $A\beta_{40}$ levels by 25% in transgenic mice, while compound concentrations in the brain were found to be approximately 1.1 μM [71].

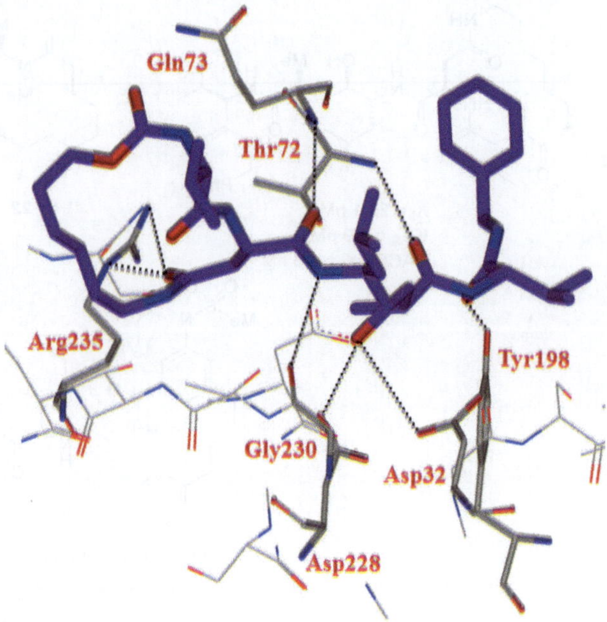

Fig. 15 X-ray crystal structure of **21**-bound BACE1 (PDB: 1XS7)

3.3 Nonpeptide Inhibitors

In general, the large molecular size and a high degree of peptidic features of peptidomimetic inhibitors led to limitations in the development of effective BACE1 inhibitor drugs. The relatively large active site of BACE1 and the need for efficient brain penetration of BACE1 inhibitors made the design task very challenging. For these reasons, the development of small-molecule and nonpeptide BACE1 inhibitors with different chemotypes has evolved. A number of high-throughput screening approaches have been pursued. These approaches have identified a wide variety of nonpeptidic scaffolds for BACE1 inhibitor design (Fig. 16). Structure-based optimization led to the development of BACE1 inhibitors that are smaller in molecular size and nonpeptidic in nature with improved pharmacokinetic properties and BBB penetration. A number of recent reviews covered this development in detail leading to the discovery of potent and selective BACE1 inhibitors [11, 72]. Herein, we plan to provide a brief outline.

3.3.1 Acyl Guanidine-Based Inhibitors

Acyl guanidine-based BACE1 inhibitors were discerned by high-throughput screening (HTS). The acyl guanidine core shown in Fig. 17 makes interactions with the catalytic aspartic acids (Asp32 and Asp228) of BACE1 [73]. As shown in Fig. 18, inhibitor **24** with an α-substituted 2,4-biphenylpyrrole showed an enzyme

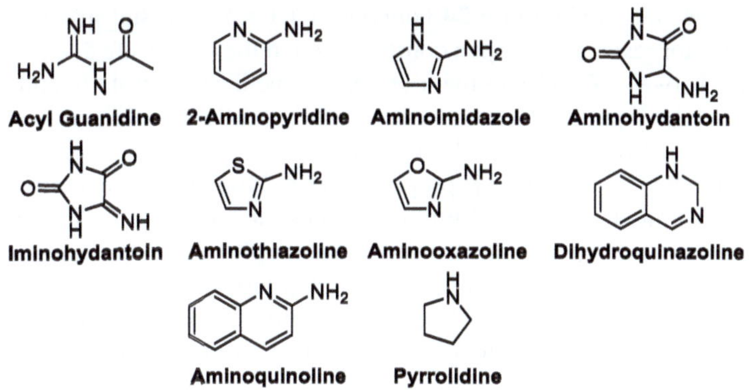

Fig. 16 Common scaffolds for nonpeptide BACE1 inhibitors

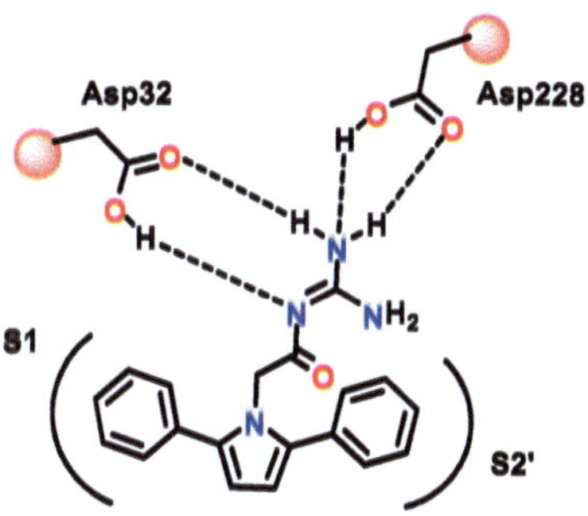

Fig. 17 Binding mode of inhibitor **21**

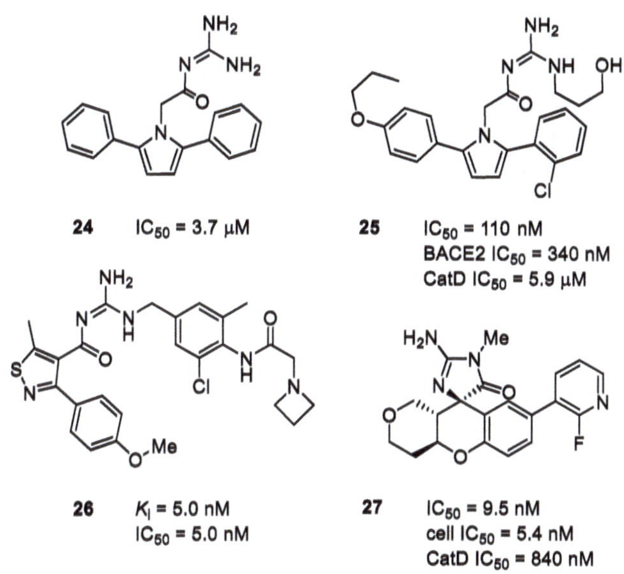

Fig. 18 Structures and activity of acyl guanidine-based inhibitors **24–27**

IC_{50} of 3.7 μM [73]. An inhibitor **24**-bound BACE1 X-ray crystal structure revealed that the inhibitor resides in an open conformation in the active site due to stabilizing interactions between Tyr71 and the π-system of the diarylpyrrole ligand (Fig. 17). Furthermore, the biphenyl ligand resides in both the S1 and S2′ sites with the P1 phenyl positioned directly toward an unoccupied S3 subsite. To optimize ligand binding in the S1–S3 regions, a *p*-propyloxyphenyl moiety was examined in place of the unsubstituted aryl ring. Similarly, access to potential hydrogen bonding interactions in the S1′ pocket was promoted through the functionalization of one of the guanidine nitrogens with a propyl alcohol group. This resulted in a substantial improvement in potency (IC_{50} 110nM) for inhibitor **25** (Fig. 18). These biaryl pyrrole-based inhibitors show low levels of selectivity over other human aspartyl proteases.

Inhibitor **26**, with an isothiazole derivative, was optimized after the discovery of the scaffold by HTS. This inhibitor exhibited excellent BACE1 affinity ($K_i = 5.0$ nM, Fig. 18) [74]. In vivo viability was evaluated with dose-dependent studies in mouse models. Plasma $A\beta_{40}$ levels were found to be diminished, while brain and CSF $A\beta_{40}$ levels were maintained during the treatment. This suggests that the inhibitor is a substrate for Pgp. However, further improvement is necessary for better BBB penetration.

A series of chromane-based spirocyclic acyl guanidine-derived BACE1 inhibitors were developed. Inhibitor **27** showed very good BACE1 and Aβ cell inhibitory activity [75]. The inhibitor displayed a high efflux ratio. However, it was able to reduce CSF $A\beta_{40}$ levels from 53% to as high as 63% over multiple species [75].

3.3.2 2-Aminopyridine-Based Inhibitors

Small-molecule nonpeptide BACE1 inhibitors have been developed based upon the 2-aminopyridine scaffold. BACE1 bound to 2-aminopyridine-based inhibitors adopts a flap-open conformation that results from the displacement of Tyr71 to a position above the pyridine scaffold. In silico screening of varying sites and degree of substitution on the pyridine ring led to the identification of many lead structures. Among the initial hits, 2,6-substituted aminopyridine **28** displayed a modest potency with an IC_{50} of 25 μM (Fig. 19) [76]. A 2,3-substituted aminopyridine derivative **29** showed BACE1 IC_{50} of 690 nM. This lead structure was then optimized for the S1–S3 hydrophobic pockets. The biaryl-substituted pyrrole substituent was incorporated to improve the cellular potency of inhibitors. Inhibitor **30** displayed an improved cellular potency ($EC_{50} = 440$ nM) (Fig. 19) [77]. Furthermore, it exhibited >100-fold selectivity against BACE2 and >500-fold selectivity against CatD. The aminopyridine moiety of inhibitor **30** binds to the catalytic aspartic acids, while the pyrimidine moiety forms a hydrogen bond with Ser229 (Fig. 20).

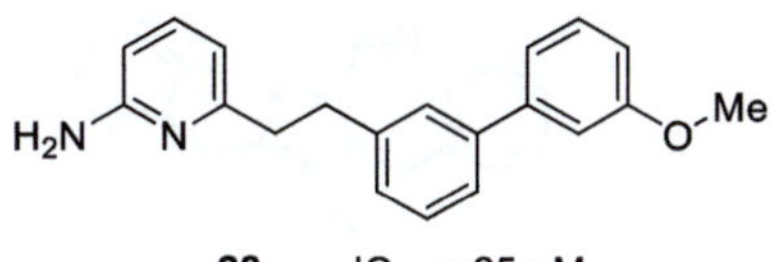

28 IC_{50} = 25 μM

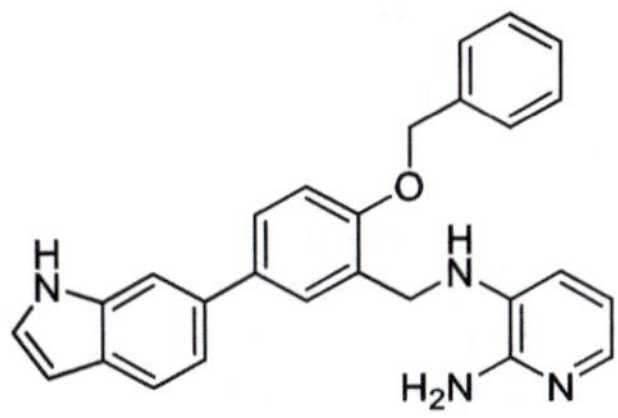

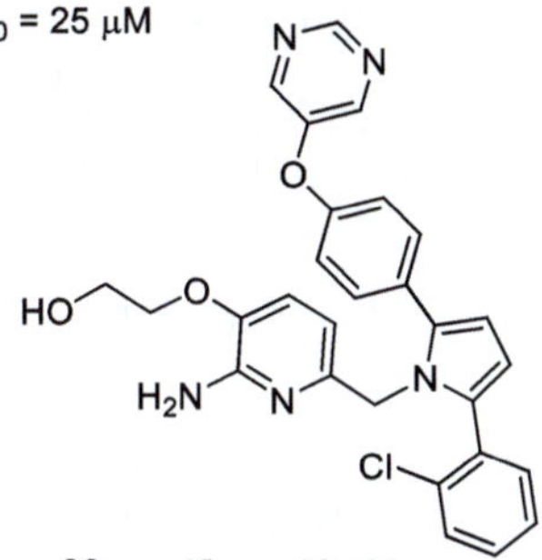

29 IC_{50} = 690 nM

30 IC_{50} = 40 nM
EC_{50} = 440 nM
BACE2 = 5.43% @ 12.5 μM
CatD = 23% @ 12.5 μM

Fig. 19 Structures and activity of 2-aminopyridine-based inhibitors **28–30**

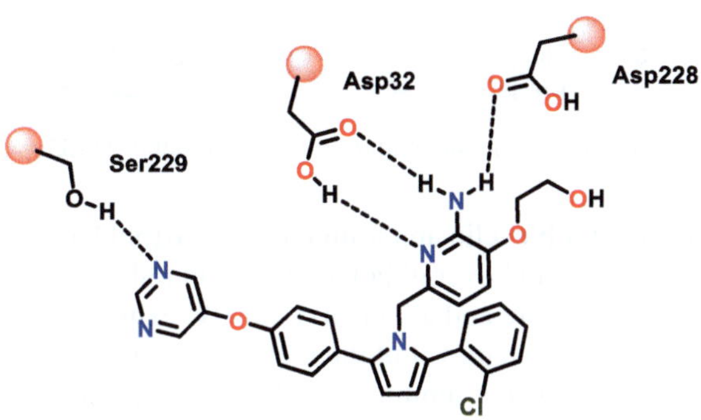

Fig. 20 Binding mode of inhibitor **30**

3.3.3 Aminoimidazole-Based Inhibitors

Potent and selective BACE1 inhibitors emerged based upon the aminoimidazole heterocyclic scaffold. As shown in Fig. 21, HTS initially yielded dibenzyl-substituted imidazole **31** as a weak inhibitor. Molecular modeling studies showed that the amino group of the imidazole heterocycle was responsible for binding with the catalytic aspartic acid residues. Subsequent optimization of the scaffold led to the design of conformationally constrained **32** with an enzymatic IC_{50} of 63 nM [78]. This increase in potency has been attributed to the additional hydrophobic interactions with the flap region of BACE1 as a result of the constrained indane moiety. This inhibitor showed a good Pgp efflux ratio of 3.6, suggesting viable brain penetration.

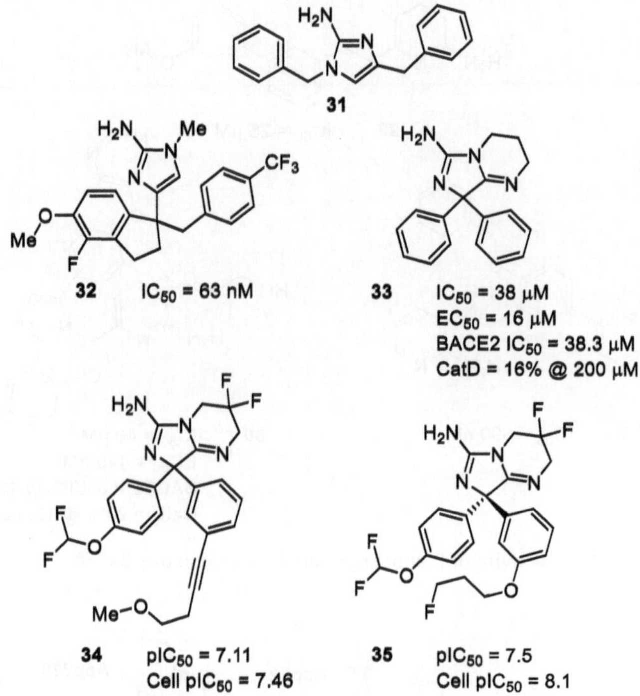

Fig. 21 Structures and activity of aminoimidazole-based inhibitors **31–35**

Further evolution of a bicyclic aminoimidazole scaffold has provided inhibitors with better brain efficacy and overall potency. The initial hit compound **33**, with a bicyclic aminoimidazole core and a biphenyl moiety, exhibited an IC_{50} in the high micromolar range [79]. The removal of the biphenyl motif and functionalization of the aryl rings resulted in potent inhibitors with better cell membrane permeability and Pgp efflux profiles. Inhibitor **34**, with a *p*-difluoromethyl ether substitution on one of the phenyl rings and an *m*-alkynyl substituent on the other ring, displayed good potency in both enzymatic and cellular assays with pIC_{50} values of 7.11 and 7.46, respectively [80]. Cell membrane permeability, as determined by a Caco-2 assay, was 8.4×10^{-6} cm/s, and the efflux ratio was 3.5, indicating potential for BBB penetration. An X-ray crystal structure of **34**-bound BACE1, shown in Fig. 22, provided insight into the ligand-binding interactions of this inhibitor [80]. The structure was determined at 2.0 Å resolution. The aminoimidazole moiety interacted with the catalytic Asp32 and Asp228 residues, binding in a flap-open conformation. This allows Trp76 to be in position for hydrogen bonding to the oxygen of the *p*-difluoromethyl ether. The alkynyl substituent of the second ring extends into the S3 pocket. Replacement of the alkyne chain of inhibitor **34** with a fluorinated propyl ether resulted in inhibitor **35**. This optically active inhibitor showed similar potency as **34** [80]. Interestingly, the efflux ratio of **35** was enhanced to a value of 0.8. Further, in vivo assessment in a mouse model using

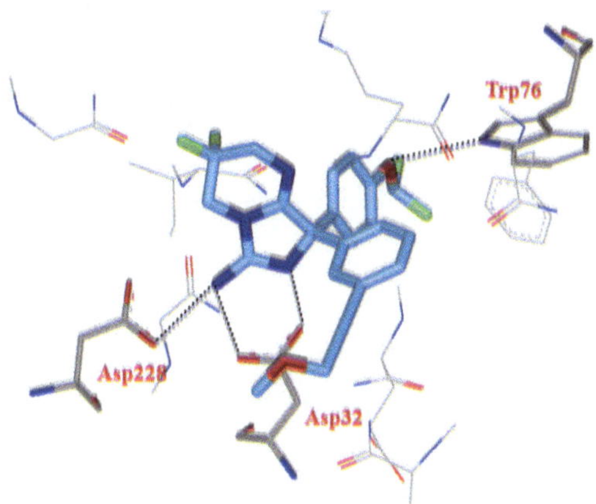

Fig. 22 X-ray crystal structure of **34** in the enzyme active site (PDB: 4ACX)

oral coadministration of **35** with a Pgp inhibitor resulted in reduction of brain and plasma Aβ levels by 17% and 76%, respectively [80].

3.3.4 Aminohydantoin-/Iminohydantoin-Based Inhibitors

Identification of initial hit compound **36** with an aminohydantoin scaffold led to the development of potent and selective BACE1 inhibitors. As shown in Fig. 23, inhibitor **36** displayed an enzymatic IC_{50} of 3.4 μM [81]. Subsequent optimization of the hydrophobic core led to inhibitor **37** with an improvement in potency. X-ray structural studies of **37**-bound BACE1 revealed that the catalytic aspartic acid residues were participating in hydrogen bonding interactions with the amino moiety and the N-H nitrogen of the pharmacophore. Furthermore, the phenylpyridine moiety extends into the S3 pocket to make an interaction with Ser229 through a water bridge. Interactions with the Trp76 within the S2′ pocket were made possible through the *p*-methoxybenzyl moiety. Inhibitor **37** exhibited good BACE1 activity with an IC_{50} of 10 nM and a cellular EC_{50} of 20 nM (Fig. 23) [81]. Furthermore, it displayed greater than 80-fold selectivity against BACE2, CatD, and other aspartic acid proteases. In vivo studies in a transgenic mouse model showed that a 100 mg/kg oral dose of inhibitor **37** cleared 69% of plasma $A\beta_{40}$ [81].

Further exploration of an aminocyclohexyl group as the S1 ligand resulted in inhibitor **38**. This inhibitor displayed a low nanomolar BACE1 activity ($IC_{50} = 22$ nM) and a 1136-fold selectivity against CatD [82]. As shown in Fig. 24, the X-ray crystal structure of **38**-bound BACE1 revealed a unique binding mode in the active site. The quinoline group unexpectedly occupies a solvent-exposed region of the active site and makes additional hydrophobic interactions with Ile110 and Lys107. Additionally, it forms a hydrogen bonding interaction with the backbone carbonyl of Phe108. The inhibitor was able to reduce plasma $A\beta_{40}$ by 65% and 55% following oral and subcutaneous administration in rats. Brain $A\beta_{40}$, however, was not reduced when dosed in a mouse model. In whole cell assays, the

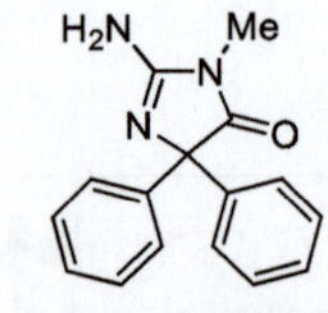

36 IC_{50} = 3.44 μM
EC_{50} = 3.4 μM
BACE2 IC_{50} = 770 nM
CatD = 36% @ 100 μM

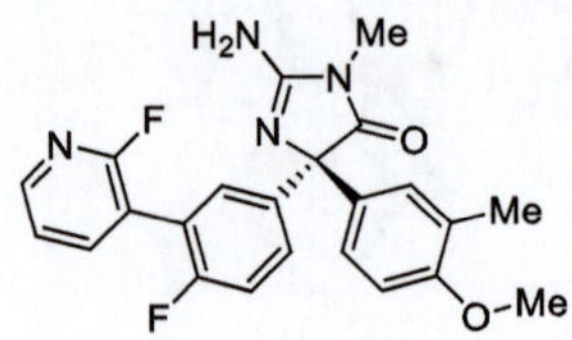

37 IC_{50} = 10 nM
EC_{50} = 20 nM
BACE2 IC_{50} = 810 nM
CatD IC_{50} = 820 nM

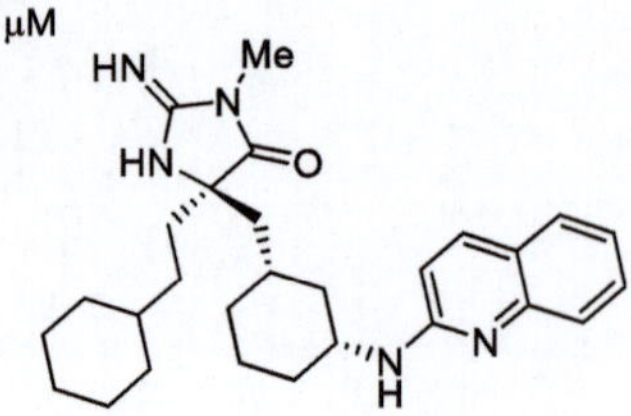

38 K_i = 22 nM
IC_{50} = 258 nM
CatD K_i = 25 μM

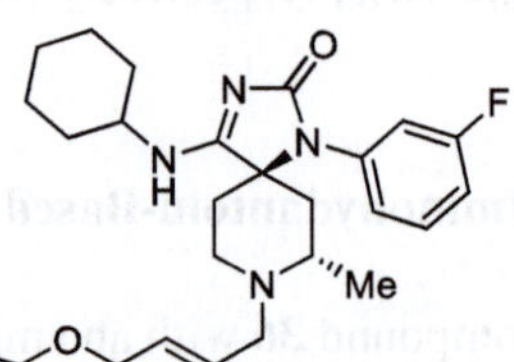

39 IC_{50} = 6 nM
Cell IC_{50} = 63 nM
BACE2 IC_{50} = 2 nM
CatD IC_{50} = 16.3 μM

Fig. 23 Structures and activity of aminohydantoin-/iminohydantoin-based inhibitors **36–39**

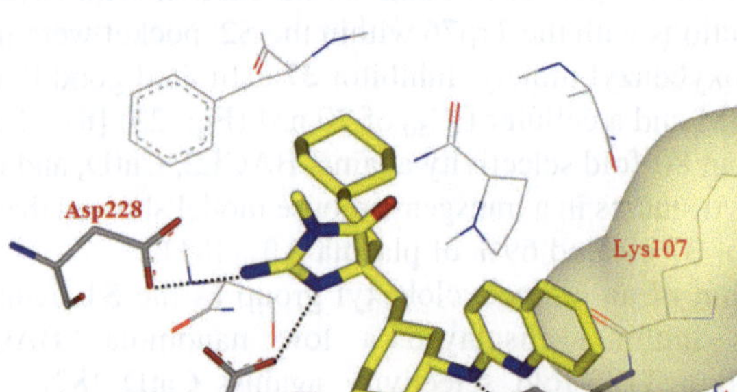

Fig. 24 X-ray crystal structure of **38**-bound BACE1 (PDB: 4R95)

inhibitor showed a cellular EC_{50} of 258 nM. This was further improved to an EC_{50} of 110 nM upon coadministration with a Pgp inhibitor [82].

The development of a closely related spirocyclic piperidine resulted in potent inhibitor **39** [83]. The X-ray crystal structural studies revealed that the binding mode of the inhibitor results in a slightly larger pocket, which can be further optimized. The inhibitor showed no selectivity against BACE2 ($IC_{50} = 2$ nM) but shows good selectivity against CatD ($IC_{50} = 16.3$ μM). It also showed improved cellular potency of 63 nM and a Pgp efflux ratio of 4.4.

3.3.5 Aminothiazoline- and Aminooxazoline-Based Inhibitors

The development of BACE1 inhibitors with an aminothiazoline scaffold was based upon the initial hit compound **40** [84]. As shown in Fig. 25, inhibitor **40** displays a modest enzymatic potency of 41.2 μM. Initial X-ray studies of **40**-bound BACE1 revealed that meta-substitution of the phenyl moiety could maximize interactions within the S3 subsite. Also, the *meta*-substituted amide group allows the aromatic ring to adopt a nearly planar conformation through a hydrogen bonding interaction with Gly291. Extensive SAR studies led to inhibitor **41**, which shows good BACE1 and cellular Aβ inhibitory activity [84]. However, inhibitor **41** turned out to be a good Pgp substrate. It shows 10% reduction of $A\beta_{40}$ levels with an oral dose of 30 mg kg^{-1}.

Inhibitors with related aminooxazoline scaffolds displayed good BACE1 inhibitory and cellular activity. Inhibitor **42** incorporates a trifluoromethyl group on the oxazoline heterocycle. It shows a BACE1 IC_{50} of 12 nM and cellular Aβ inhibitory activity of 2 nM [85]. The inhibitor showed good selectivity against other common aspartic acid proteases with IC_{50} levels greater than 200 μM. In vivo studies of

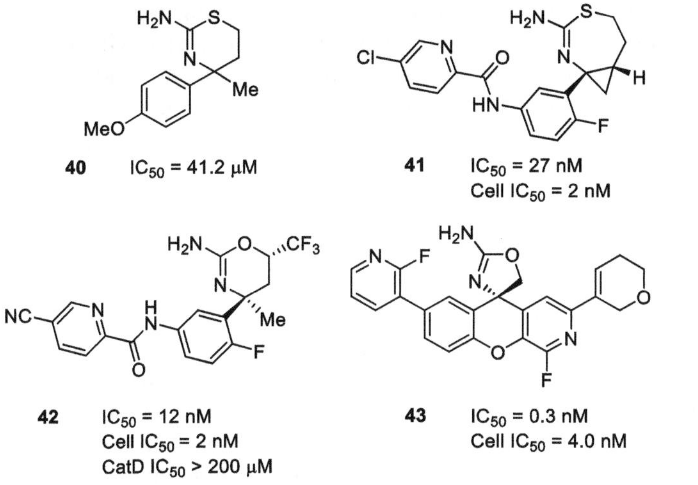

40 IC_{50} = 41.2 μM

41 IC_{50} = 27 nM
Cell IC_{50} = 2 nM

42 IC_{50} = 12 nM
Cell IC_{50} = 2 nM
CatD IC_{50} > 200 μM

43 IC_{50} = 0.3 nM
Cell IC_{50} = 4.0 nM

Fig. 25 Structures and activity of aminooxazoline- and aminothiazoline-based inhibitors **40–43**

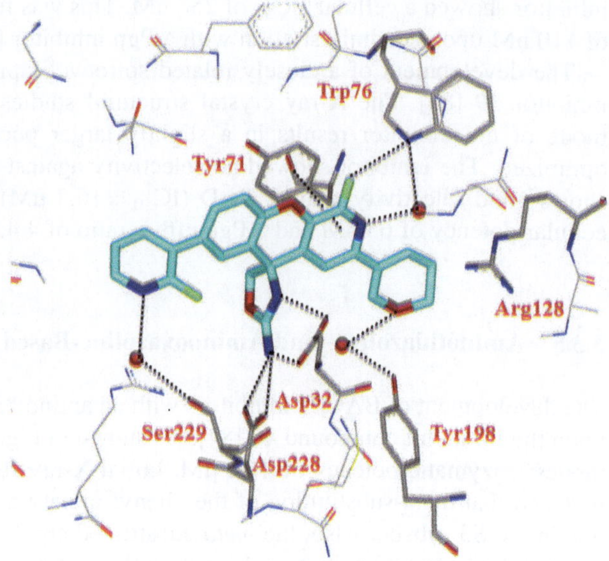

Fig. 26 X-ray crystal structure of **43** in the BACE1 active site (PDB: 4WTU)

inhibitor **42** with an oral dose as small as 1 mg/kg showed reduction of $A\beta_{40/42}$ levels for up to 24 h. It also displayed a favorable pharmacokinetic profile [85].

Small-molecule BACE1 inhibitors including a xanthene core with an amino-oxazoline head group were explored. Xanthene derivatives generally exhibited good potency but showed high efflux activity. Further modification of the xanthene core and its substituents were pursued in order to develop drug-like inhibitors [86]. Inhibitor **43** with a 3-aza-2-fluoroxanthene scaffold showed excellent BACE1 and cell inhibitory potency [87]. As shown in Fig. 26, an X-ray crystal structure of inhibitor **43**-bound BACE1 revealed that the 3-azaxanthene core formed a crucial hydrogen bonding interaction with Trp76. The oxygen of the dihydropyran also formed a hydrogen bonding interaction with Tyr198. The nitrogen of the pyridine engaged in an additional hydrogen bonding interaction with Ser229. Inhibitor **43** was orally administered to a number of species and showed a robust reduction of central nervous system (CNS) $A\beta_{40}$ levels [87].

3.3.6 Dihydroquinazoline-Based Inhibitors

BACE1 inhibitors containing dihydroquinazoline scaffolds were developed after identification of fragment hit **44** (Fig. 27) [88]. The X-ray structure of **44**-bound BACE1 shows that the inhibitor adopts a hairpin conformation (Fig. 28), enabling the inhibitor to occupy the subsite specific S1 pocket. The X-ray crystal structure reveals a vacant hydrophobic S1′ pocket available for optimization of ligand binding. Incorporation of a cyclohexyl group with (*S*)-configuration led to inhibitor **45** [88]. Presumably, the cyclohexyl ligand fills in the S1′ hydrophobic pocket. As a result, this inhibitor showed an enzymatic Ki_{50} of 11 nM. The inhibitor displayed

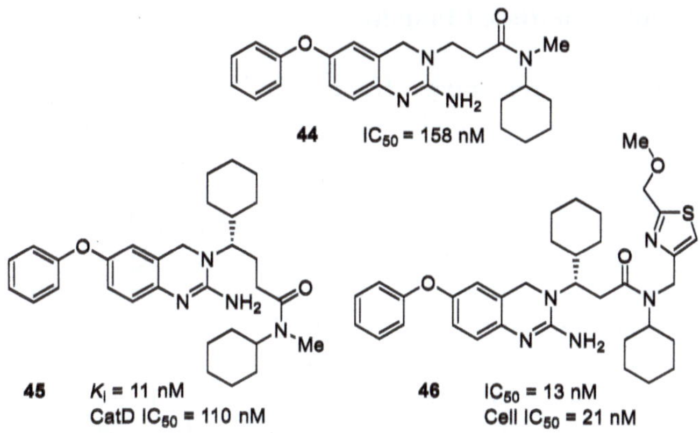

Fig. 27 Structures and activity of dihydroquinazoline-based inhibitors **44–46**

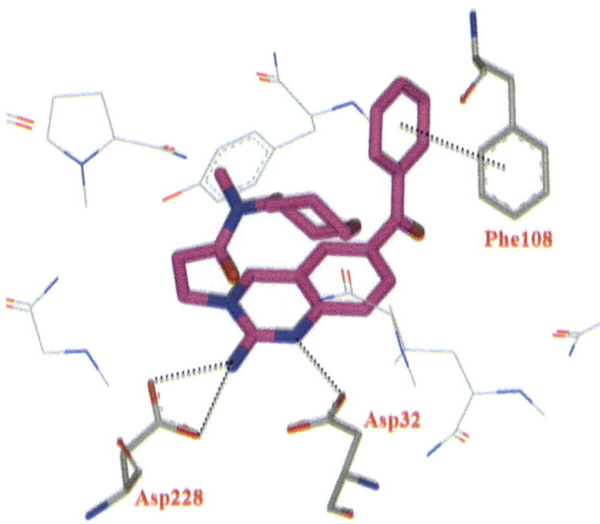

Fig. 28 X-ray crystal structure of **44**-bound BACE1 (PDB: 2Q11)

modest selectivity against common aspartyl proteases such as renin ($IC_{50} = 2.7$ μM) and CatD ($IC_{50} = 0.11$ μM). Oral administration of inhibitor **45** in rats (30 mg/kg) exhibited as much as 70% reduction of $A\beta_{40}$ plasma levels. However, an efflux ratio of 12.2 suggests poor brain permeability for this inhibitor. Further optimization of the ligand binding led to the design of inhibitor **46** [89]. In this inhibitor, a methoxymethyl thiazole was incorporated in place of the methyl amide. Inhibitor **46** exhibited good BACE1 enzymatic inhibitory activity ($IC_{50} = 13$ nM) and cellular Aβ inhibitory activity ($IC_{50} = 21$ nM).

3.3.7 Aminoquinoline-Based Inhibitors

Initial fragment-based screening efforts identified aminoquinoline **47** with BACE1 K_d of 900 μM [90]. Initial SAR studies showed that incorporation of an aryl ring at the 6 position of the aminoquinoline may lead to improved potency. Further structural modifications with the introduction of a *N*-cyclohexyl propionamide at C3 of the quinoline ring led to inhibitor **48** (Fig. 29) [90]. This inhibitor showed good BACE1 IC_{50}. Further structural studies suggest that a longer and bulkier alkyl chain could potentially extend further into the S2′ subsite. Inhibitor **49** incorporated an (*R*)-α-methyl on the side chain and a chloropyridyl group to improve pharmacokinetic properties [90]. Inhibitor **49** displayed good activity with a BACE1 IC_{50} of 11 nM and cellular IC_{50} value of 80 nM. It also exhibited an efflux ratio of 3.1. In vivo studies with a 60 mg/kg dose administered subcutaneously in rats showed reduction of $A\beta_{40/42}$ CSF levels by 42% postinjection. However, inhibitor **49** showed rapid clearance from human and rat microsomes. The X-ray structural studies with **49**-BACE1 complex showed that the aminoquinoline formed a tight interaction with flap residue Tyr71, which orients the quinoline and amino nitrogens toward the catalytic Asp32 and Asp228 residues (Fig. 30) [90].

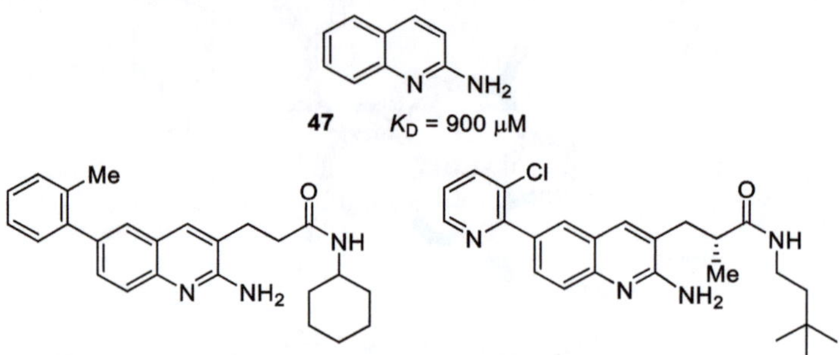

47 K_D = 900 μM

48 IC_{50} = 74 nM

49 IC_{50} = 11 nM
Cell IC_{50} = 80 nM

Fig. 29 Structures and activity of aminoquinoline-based inhibitors **47–49**

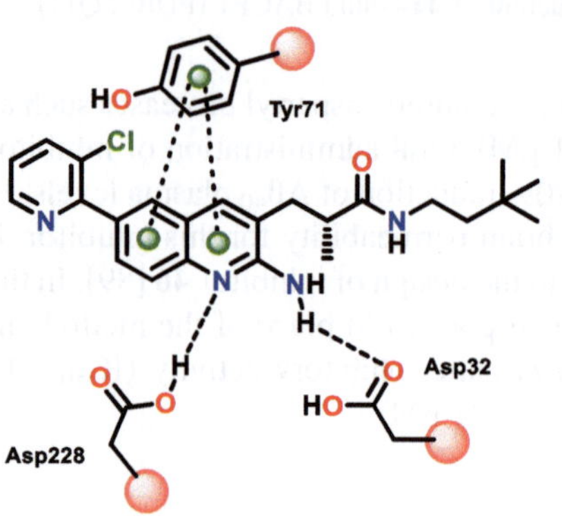

Fig. 30 Binding mode of inhibitor **49**

3.3.8 Pyrrolidine-Based Inhibitors

BACE1 inhibitors containing a pyrrolidine scaffold were identified by HTS. As shown in Fig. 31, initial compound **50** showed a low inhibitory activity of 240 μM [91]. SAR studies and further optimization resulted in inhibitor **51,** which showed BACE1 inhibitory activity of 29 nM. Its cellular activity was 570 nM (IC_{50}), and the inhibitor displayed no selectivity against BACE2 but >200-fold selectivity over CatD [91]. An inhibitor-bound X-ray crystal structure revealed that the pyrrolidine scaffold formed two hydrogen bonds with the catalytic aspartic acid residues. The *cis*-2,4-substituted piperidine ligand nicely occupied a pocket below the flap region (Fig. 32).

3.3.9 Macrocyclic Nonpeptide Inhibitors

The development of macrocyclic nonpeptide inhibitors was carried out in an effort to improve the inhibitors' properties. As mentioned earlier, the dihydroquinazoline-based inhibitors adopt a tight hairpin conformation [88]. A macrocyclic inhibitor may provide stabilization of the bioactive hairpin conformation, leading to an enhancement in potency. As shown in Fig. 33, macrocyclic inhibitor **52** displayed a BACE1 K_i of 5 nM [92]. The cellular Aβ inhibitory activity was also improved to 7 nM (IC_{50}) compared to the acyclic inhibitor. Further evaluation revealed that

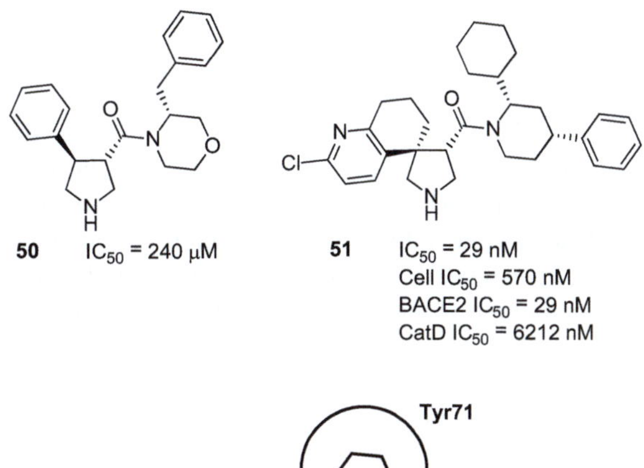

50 IC_{50} = 240 μM

51 IC_{50} = 29 nM
Cell IC_{50} = 570 nM
BACE2 IC_{50} = 29 nM
CatD IC_{50} = 6212 nM

Fig. 31 Structures and activity of pyrrolidine-based inhibitors **50** and **51**

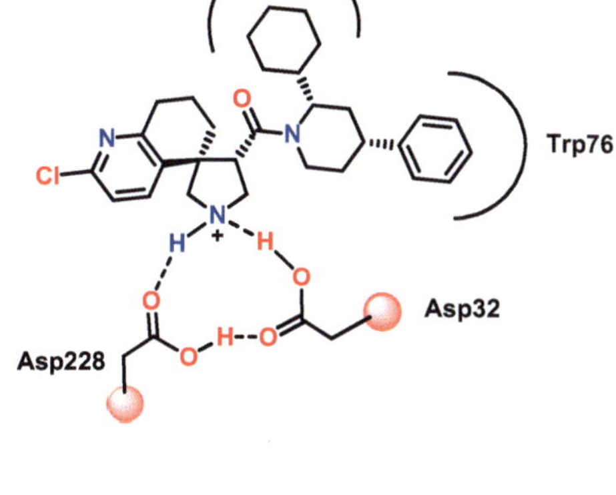

Fig. 32 Binding mode of inhibitor **51**

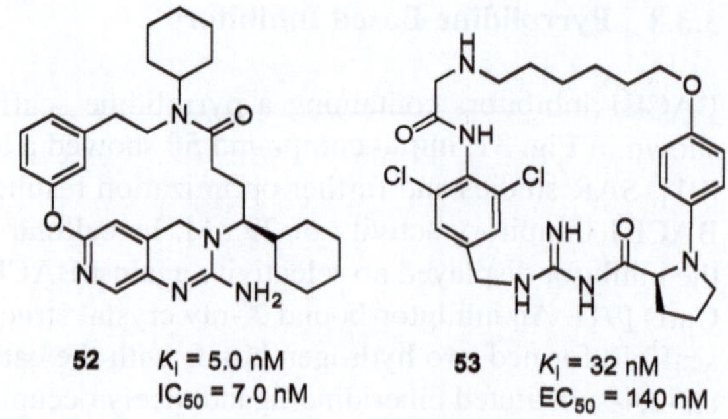

Fig. 33 Structures and activity of nonpeptidic macrocyclic-based inhibitors **52** and **53**

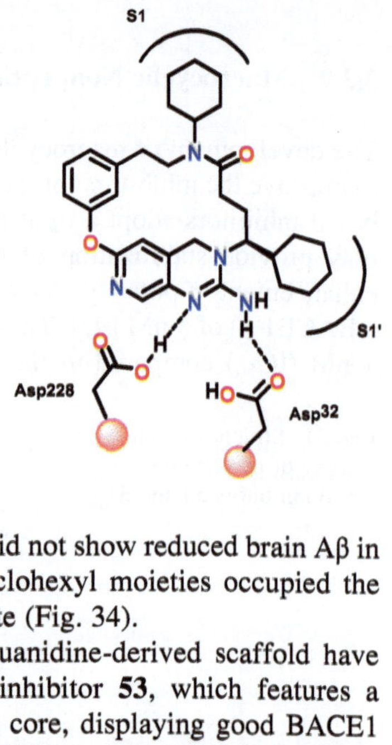

Fig. 34 Binding mode of inhibitor **52**

inhibitor **52** is a Pgp substrate and in vivo studies did not show reduced brain Aβ in animals. Structural studies suggested that the cyclohexyl moieties occupied the vacant S1 and S1′ pockets of the BACE1 active site (Fig. 34).

Macrocyclic inhibitors based upon the acyl guanidine-derived scaffold have been investigated. A representative example is inhibitor **53**, which features a 26-membered macrocycle with an acyl guanidine core, displaying good BACE1 ($IC_{50} = 32$ nM) and cellular Aβ inhibitory activity ($EC_{50} = 140$ nM) [93]. Inhibitor **53** also showed good selectivity against multiple aspartic acid proteases, including 700-fold selectivity against CatD. Furthermore, the inhibitor showed a diminished efflux ratio of 2.3. In vivo studies with wild-type mice involving subcutaneous dosing of 40 mg/kg showed a 74% reduction of plasma Aβ levels and a peripheral drug concentration of 3.8 μM 5 h postinjection. However, Aβ brain levels were found to be nearly identical to the control levels [93].

4 Clinical Evaluation of BACE1 Inhibitors

The discovery of a clinically translatable BACE1 inhibitor has proven to be a difficult challenge for the medicinal chemistry community. While many highly potent inhibitors have been synthesized and evaluated preclinically, there are many more barriers an inhibitor must cross to be considered clinically viable. Further, a drug intended to treat a brain disorder experiences additional hurdles it must overcome. First, an inhibitor must have the ability to penetrate the BBB. This requires a proper balance of hydrophobicity and hydrophilicity. Also, upon crossing the BBB, Pgp efflux can eject the inhibitor, precluding any therapeutic effect. This efflux can be avoided by carefully designing compounds with minimal recognition by Pgp.

Additionally, due to the chronic nature of AD, any approved therapy would be a maintenance drug, which patients must take for the remainder of their lives. Because of this, accumulation of the drug in body tissues and subsequent toxicity must be monitored and addressed. Further, selectivity over non-pathological aspartic acid proteases is critical, especially in the case of chronic therapy. Selectivities over systemic enzymes such as BACE2 and CatD provide an additional challenge due to the similarity of the proteolytic mechanism and substrate specificity. Unique pockets in each enzyme must be exploited to provide this necessary selectivity.

A final challenge is the balance of inhibitor size. The enzyme has a rather large active site. It has been a difficult challenge to develop compounds large enough to fill the active site and bind efficiently, yet small enough to cross the BBB, cell membrane, and endosomal membrane to inhibit BACE1 at the vesicle. Generally, large peptidomimetic inhibitors tend to have low oral bioavailability, low half-lives, and poor BBB penetration. However, small inhibitors may not adequately fill the enzymatic active site, resulting in poor affinity for the enzyme. Despite these challenges, potential inhibitors have been pursued in clinical trials (Fig. 36). Thirteen drugs have been brought to clinical trials in the United States, six of which have progressed into phase II or III clinical trials and five are currently being tested in the clinic (Figs. 35 and 36)

4.1 *Clinical Evaluation of CTS21166*

The structure of CTS21166 has not been publically disclosed. This compound, developed by CoMentis, was the first BACE1 inhibitor to be brought to clinical trials. The 8-month phase I trial began in June 2007 and tested CTS21166 in 56 healthy adult males to evaluate the safety profile of single-ascending doses as well as the pharmacokinetics and major plasma metabolites [94]. Intravenous injections of 7.5–225 mg were well tolerated and showed gradual clearance amenable to once-daily dosing across all doses. At 225 mg, plasma Aβ levels were

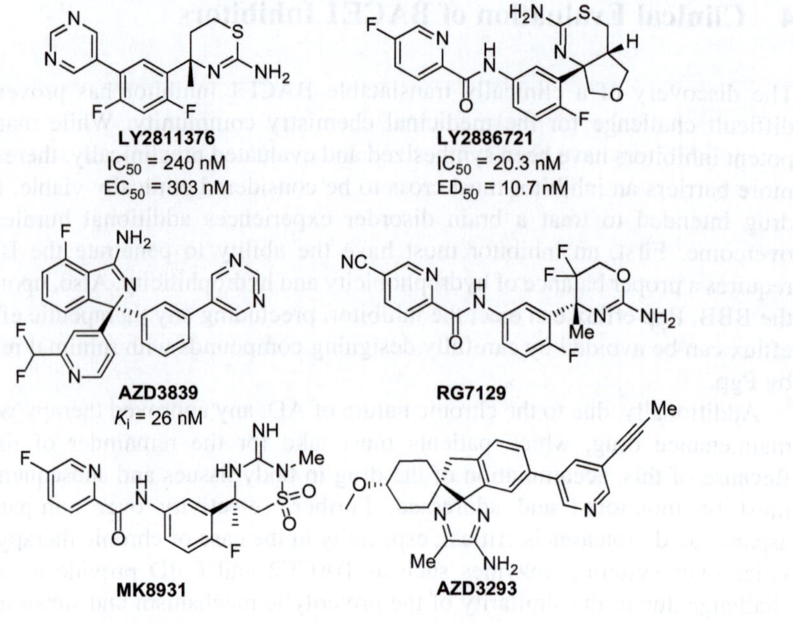

Fig. 35 Known structures of BACE1 clinical drug candidates

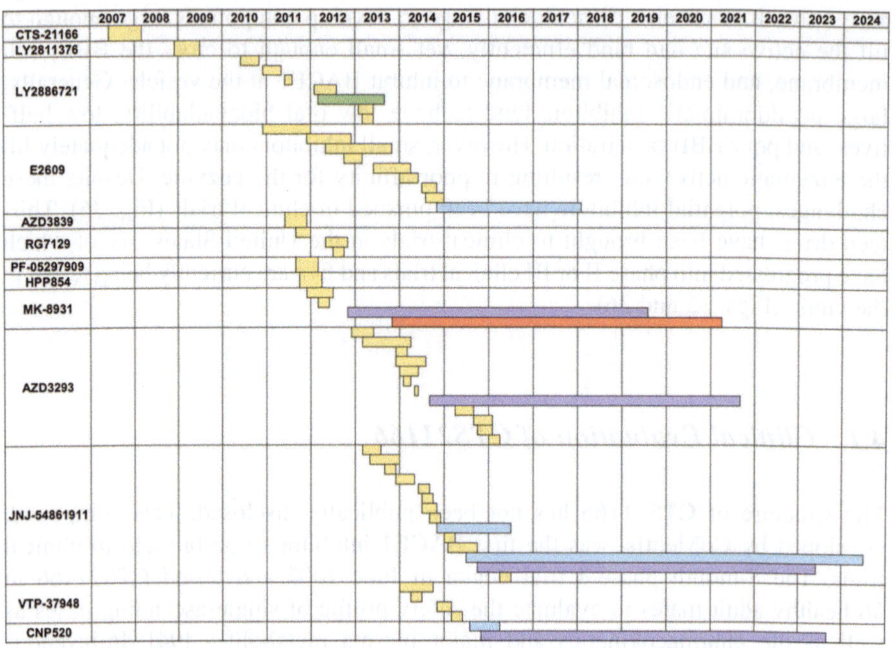

Fig. 36 Timeline of BACE1 inhibitor clinical development. Phase I = *yellow*; phase I/II = *green*; phase II = *blue*; phase II/III = *purple*; phase III = *red*

reduced by up to 80% at 3 h postinjection versus placebo [95, 96]. While CTS21166 passed phase I trials, its clinical development was not continued.

4.2 Clinical Evaluation of LY2811376

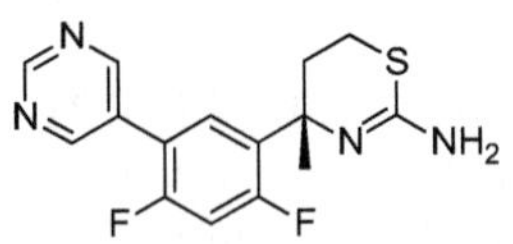

LY2811376
IC_{50} = 240 nM
EC_{50} = 303 nM

LY2811376, Eli Lilly's first clinical BACE1 inhibitor, began phase I clinical trials in 2009. It is a part of the aminothiazoline class of nonpeptidic inhibitors. The inhibitor was given to 61 healthy men and women of non-child-bearing potential to investigate the safety and tolerability at single doses from 5 to 500 mg as oral capsules [97]. LY2811376 was well tolerated with no serious adverse effects reported. Maximum plasma concentrations were reached 2 h post-dose with an average half-life of 40 h. Cerebrospinal fluid (CSF) concentrations reached maximal levels at about 5 h post-dose [98]. Dose-dependent reduction of $A\beta_{40}$ and $A\beta_{42}$ was observed in both the plasma and CSF. At 90 mg, an 80% reduction in $A\beta_{40}$ was observed at 7 h after administration, with no full recovery to pre-dose concentrations within 120 h after dosing. In the CSF, up to a 54% decrease in Aβ was observed with a statistically significant decrease being observed 7 h after dosing and a maximum reduction at 12–14 h post-dose [98].

In preparation for phase II studies, a rat toxicology study was performed alongside phase I clinical studies. During the 3-month preclinical work, it was observed that autofluorescent granules were accumulating in the retinal epithelium, neurons, and glial cells at doses greater than 30 mg/kg. Based on this data, clinical trials of LY2811376 were immediately terminated. It was hypothesized that the retinal toxicity may be due to off-target effects against other aspartic acid proteases such as BACE2 or CatD. Fortunately, no clinically significant retinal pathology was present in any of the clinical trial participants in follow-up exams [98].

4.3 Clinical Evaluation of LY2886721

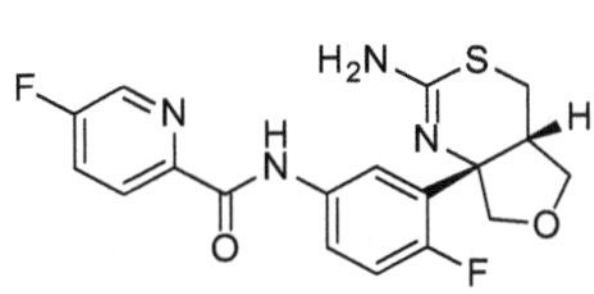

LY2886721
IC_{50} = 20.3 nM
ED_{50} = 10.7 nM

It was more than a year before the next clinical candidate was tested in human subjects. LY2886721, developed by Eli Lilly and Company, became the first BACE1 inhibitor to reach phase II trials [99]. Like LY2811376, it is a part of the aminothiazoline class of nonpeptide inhibitors. A total of six phase I studies tested the safety and pharmacology of LY2886721 in 155 subjects. Trials began in June

2010 with a single-ascending dose study in healthy adult males and females, administered as oral capsules in variable doses ranging from 1 to 200 mg or placebo. Concentrations of LY2886721 and $A\beta_{40}$ were measured up to 8 days post-dose in the plasma and up to 36 h post-dose in the CSF [100]. Another study investigated the tolerability of multiple-ascending doses of 5–35 mg or placebo administered as oral capsules in 42 healthy adults. Subjects were monitored for clinically significant effects as well as concentrations of LY2886721 and $A\beta_{40}$ in the plasma and CSF [101].

An open-label study with radioactive ^{14}C-LY2886721 in eight healthy males was used to determine metabolism and excretion pathways [102]. An additional single- and multiple-dose study was performed on healthy volunteers before proceeding to phase I/II studies in Alzheimer's patients [103]. LY2886721 was evaluated in 128 patients with mild cognitive impairment or mild Alzheimer's disease. Doses of 15, 35, or 70 mg or placebo were administered orally once daily for 26 weeks. Concentrations of $A\beta_{40}$ and $A\beta_{42}$ were measured at 12 and 26 weeks in both the CSF and plasma. Other cognitive tests were performed at 26 weeks including a Neuropsychological Test Battery (NTB), Alzheimer's Disease Assessment Scale (ADAS), Clinical Dementia Rating (CDR) Scale, and Mini Mental State Examination (MMSE) [104]. Concurrently, additional phase I studies were performed to evaluate two formulations of LY2886721 and assess safety and side effects in Alzheimer's patients and high-dose tolerability in healthy volunteers [105]. However, in June 2013, Lilly voluntarily halted the phase II trial and all further clinical development of LY2886721 due to abnormal liver biochemistry observed in subjects. It is believed that the abnormalities are not related to the BACE1 mechanism [106]. No clinical trial results of LY2886721 have been published.

4.4 Clinical Evaluation of E2609

Eisai Inc. brought clinical candidate E2609 to trial, starting with an initial phase I safety study beginning in December 2010. The structure of E2609 has not been publically disclosed. Unpublished preclinical work showed that E2609 lowered Aβ levels in the CSF and plasma of rats, guinea pigs, and non-human primates [107]. In the first single-ascending dose study, E2609 was dosed orally in healthy adult volunteers up to 800 mg or placebo and in healthy elderly volunteers at 50 mg or placebo. Subjects were monitored for adverse effects, and pharmacokinetic parameters were assessed in plasma and urine [108]. Adverse effects were reported in 21.8% of subjects treated with E2609 compared to 16.7% dosed with placebo. No serious adverse effects were reported, demonstrating that E2609 was well tolerated in all dosing groups. Plasma concentration of E2609 was dose dependent, with a half-life of 15.9 h at 800 mg. Maximal reduction of plasma Aβ occurred from 6 to

24 h after administration with decreases ranging from 56.3% at 5 mg to 91.0% at 800 mg. At up to 144 h post-dose, an 800 mg dose still afforded a 45% reduction of Aβ in the plasma [109].

During a multiple-ascending dose study, healthy adult volunteers received 25–400 mg or placebo administered orally once daily for 14 days. Subjects were monitored for adverse effects, as well as Aβ concentration in the plasma and CSF [110]. E2609 continued to be well tolerated in multiple doses. Plasma levels of E2609 were dose dependent, with a constant concentration achieved at 7 days. Concentration of Aβ in the plasma was significant, with reduction greater than 85% at the lowest dose. Assessment of CSF showed that at 100 mg, $A\beta_{40}$ and sAPPβ were reduced by 82.1% and 94.2%, respectively, while sAPPα levels increased by 128.6%, indicating enhanced processing by α-secretase due to the inhibition of BACE1 [111].

Other phase I studies included a safety and pharmacodynamic study in 65 mild AD patients given a single oral dose ranging from 5 mg to 400 mg [112] and a bioavailability and absorption assessment in 18 healthy males [113]. Utilizing ^{14}C-radiolabeled compound, the metabolism and excretion pathway of E2609 was investigated in 113 healthy males at a single 100 mg oral dose [114]. To investigate potential drug-drug interactions, 195 healthy adults were dosed with E2609 in the presence or absence of itraconazole, an antifungal and cytochrome P450 3A4 inhibitor; rifampin, an antibiotic and hepatic cytochrome P450 inducer; digoxin, a treatment for heart conditions; and donepezil, an acetylcholinesterase inhibitor used for palliative care in Alzheimer's patients [115]. To assess pharmacokinetic parameters, a single oral dose study in Japanese and white healthy males was performed monitoring plasma concentrations and parameters for up to 10 days [116]. Another drug-drug interaction study was performed on 60 healthy adult volunteers to investigate the effect of moxifloxacin with E2609. Moxifloxacin is an antibiotic which has shown to prolong the QT interval when dosed with some other medications. In this study, particular attention was given to changes in the QTc interval and ECG recordings [117].

In November 2014, Eisai Inc., in collaboration with Biogen, began a phase II study of E2609 in patients with early AD and mild-to-moderate dementia due to AD. An estimated 700 patients will be administered a low, middle, or high dose of E2609 or placebo orally once per day for 18 months. End points will be evaluated via change in derived Alzheimer's Disease Composite Score (ADCS) at 18 months, and change in hippocampal atrophy as determined by MRI, and CSF Aβ levels [118]. Clinical evaluation of this compound is ongoing, and the current phase II study is estimated to be completed in January 2018.

4.5 Clinical Evaluation of AZD3839

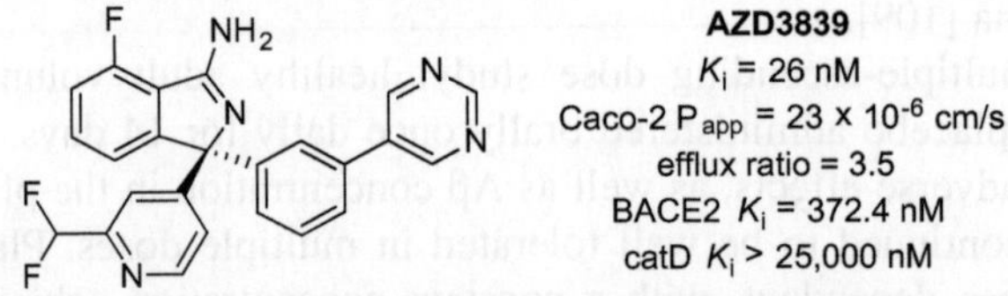

In June 2011, AstraZeneca brought clinical candidate AZD3839 into phase I trials. AZD3839 was developed from an aminopyrimidinone NMR hit. Subsequent structural optimization led to the aminoisoindole derivative that was ultimately assessed in the clinic [119]. In preclinical studies, AZD3839 displayed an enzymatic K_i of 26.1 nM/L and a 14-fold and 1,000-fold selectivity over BACE2 and CatD, respectively [119]. In mice, the brain concentration of AZD3839 reached maximum concentration at 30 min and returned to baseline levels at 4.5 h when dosed at 80 μmol/kg. When the dose was doubled, the brain exposure elongated to about 8 h. $A\beta_{40}$ levels decreased 30% and 50%, respectively [119]. Guinea pigs were given an oral dose of either 100 or 200 μmol/kg. Lasting up to 8 h, guinea pigs which received 200 μmol/kg had 20–60% lower brain $A\beta_{40}$ levels than the control. The lower-dose group showed a brain $A\beta_{40}$ reduction of 20–30% after 1.5–4.5 h post-dose. CSF $A\beta_{40}$ levels in the high-dose group were reduced by 50% at 3 h and 40% at 8 h; however, the reduction was not statistically significant. Plasma $A\beta_{40}$ levels were reduced by 30–80% for both dose classes over the entire test period [119]. In a non-human primate model using cynomolgus monkeys, an intravenous 20 μmol/kg dose of AZD3839 reduced CSF levels of $A\beta_{40}$, $A\beta_{42}$, and sAPPβ between 3 and 12 h of exposure. A lower dose of 5.5 μmol/kg did not show any significant effect on biomarker levels. This is proposedly due to low exposure at this concentration, as CSF concentration of AZD3839 peaked at 15 min and decreased tenfold at 3 h [119].

Based on this preclinical data, AZD3839 was brought to phase I clinical studies. A single-dose safety and tolerability study was performed in 72 healthy male volunteers given oral doses ranging from 1 to 300 mg or placebo. Subjects were monitored for adverse effects, as well as for plasma AZD3839, $A\beta_{40}$, and $A\beta_{42}$ levels [120]. AZD3839 was well tolerated at all doses, with 31% of the test group reporting at least one mild adverse event, compared to 39% in the placebo group. A dose-dependent elongation of the QT interval was observed, with an average prolongation of 20 ms at the highest dose of 300 mg. AZD3839 lowered plasma $A\beta_{40}$ and $A\beta_{42}$ in a dose-dependent manner. The maximum reduction of plasma $A\beta_{40}$ and $A\beta_{42}$ was 56% and 39%, respectively [121]. This phase I study was completed in November 2011, and AZD3839 was not brought forward into further clinical studies.

4.6 Clinical Evaluation of RG7129

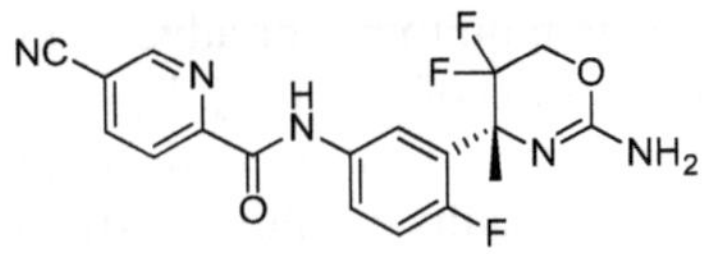

RG7129
BACE1 IC_{50} = 30 nM
HEK293APP cell IC_{50} = 18 nM
BACE2 IC_{50} = 40 nM

Hoffmann-La Roche brought inhibitor RG7129 (RO5508887) to clinical trials in September 2011. It is a member of the aminooxazoline class of nonpeptidic inhibitors. Preclinical evaluation showed that RG7129 was potent against BACE1 (IC_{50} = 30 nM) but was not selective over BACE2 [122]. Single-ascending oral doses were administered to 49 healthy male volunteers. Plasma and urine levels of RG7129 were monitored from 1.5 to 144 h post-dose. Plasma levels of $A\beta_{40}$ and $A\beta_{42}$ were measured in the plasma over the same time frame. The effect of food on the pharmacokinetics of RG7129 and incident of adverse events were also monitored [123]. A secondary study evaluated biomarker levels in the plasma and CSF as well as RG7129 concentrations and adverse effects, in 42 healthy male subjects [124]. A third multiple-ascending dose phase I trial investigated the safety and tolerability of RG7129 over prolonged exposure. Subjects were monitored for the occurrence of adverse effects as well as plasma levels of biomarkers and RG7129 [125]. Results of these studies have not been published or disclosed. In October 2013, Hoffmann-La Roche ended the clinical development of RG7129 but did not provide a clinical explanation for its cessation.

4.7 Clinical Evaluation of PF-05297909

Pfizer brought its first small-molecule BACE1 inhibitor to clinical trials in September 2011; the structure has not been released. PF-05297909 was administered orally as a single dose to 51 healthy volunteers. A thorough pharmacokinetic study was of interest; plasma and CSF levels of biomarkers, in vivo IC_{50} estimation, half-life, and CSF concentration of PF-05297909 were determined [126]. At all doses, 25–325 mg, PF-05297909 was well tolerated and displayed sufficient plasma concentrations and Aβ reduction. Unfortunately, this plasma exposure and reduction was not translated to the CSF, and no significant pharmacokinetic or pharmacodynamic effect was observed [127]. Clinical development of PF-05297909 was halted in light of these phase I results.

4.8 Clinical Evaluation of HPP854

High Point Pharmaceuticals developed inhibitor HPP854, which was determined to have significant preclinical activity worthy of translation into a clinical setting.

Phase I trials of HPP854 began in October 2011. HPP854 was administered orally once daily over 28 days in ascending doses to seven patients with mild cognitive impairment or mild AD. Patients were monitored for adverse events, plasma concentration of HPP854, and CSF concentration of Aβ [128]. Upon the conclusion of the phase I study in March 2012, High Point Pharmaceuticals terminated the clinical development of HPP854. No clinical results or inhibitor structure has been released.

4.9 Clinical Evaluation of MK-8931

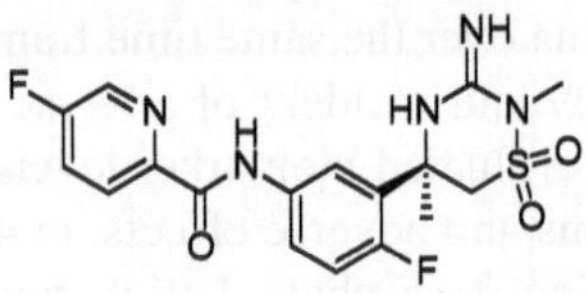

MK-8931

MK-8931 (verubecestat) was developed by Merck. It was first brought to clinical trials in the United States in December 2011. Initial clinical studies conducted in healthy human patients consisted of a single-ascending dose and multiple-ascending dose study. Volunteers were given a placebo, a single dose of 20–550 mg, or 10–250 mg once daily for 14 days. Subjects were monitored for adverse effects and CSF biomarker concentrations. MK-8931 was well tolerated across all dosing levels in both single and multiple dosing cohorts. CSF $A\beta_{40}$ levels were reduced by 25–61% in a dose-dependent manner in the single-dose study; the multiple-dose cohort showed a dose-dependent reduction of $A\beta_{40}$ ranging from 32 to 94% on day 14 [129]. Exposure of MK-8931 was dose dependent in plasma and CSF with maximum plasma concentrations at 1–4.5 h and maximum CSF concentrations at 4–6 h with a half-life of 14–22 h [130].

A phase I study of 32 patients with mild-to-moderate AD was performed to evaluate the safety and pharmacokinetics of MK-8931 in diseased patients. Participants received a 12, 40, or 60 mg oral dose of MK-8931 or placebo once daily for 7 days and were monitored for $A\beta_{40}$ and sAPPβ levels in the CSF [131]. An additional study in renal impaired patients evaluated whether renal disease changed the pharmacokinetics of MK-8931. Patients with mild, moderate, or severe renal impairment as well as healthy control groups were given a single oral dose of MK-8931 and monitored for plasma concentration of MK-8931 over 120 h, renal clearance of MK-8931, and excretion in urine [132].

MK-8931 was subsequently brought into a two-part phase II/III trial. The first part consisted of a long-term safety study in which approximately 400 mild-to-moderate Alzheimer's patients were dosed orally, once daily with 12 or 40 mg for 78 weeks or 60 mg for 13 weeks and then 40 mg for the remaining 65 weeks. Efficacy was measured by change in Alzheimer's Disease Assessment Scale-Cognitive Subscale (ADAS-Cog) score and Alzheimer's Disease Cooperative Study-Activities of Daily Living (ADCS-ADL) score. Secondary outcome measurements included change in Clinical Dementia Rating Sum of Boxes (CDR-SB) score, Neuropsychiatric Inventory (NPI) score, MMSE score, hippocampal volume, CSF tau concentration, and brain amyloid concentration. Contingent upon the first safety trial, patients would receive the 12 or 40 mg dosages, once daily by mouth up to 260 weeks with the same primary and secondary end point examinations [133].

In December 2013, it was announced that the Data Monitoring Committee recommended continuation into part two of the phase II/III trials based on the data collected during the safety phase of the study [134]. This study is expected to conclude in July 2019. Concurrently, Merck announced the beginning of a phase III study in early AD patients. Patients will receive 12 or 40 mg MK-8931 once daily for 104 weeks. Upon satisfactory completion of part one, a second part will continue the trial for an additional 260 weeks at 12 or 40 mg. Efficacy will be measured via change in 3-domain Composite Cognition Score (CCS-3D), ADCS-ADL score, hippocampal volume, time to progression of probable AD dementia, CSF tau, and composite cortical amyloid standard uptake value ratio as determined by PET imaging [135]. This trial is currently recruiting participants, with an estimated enrolment of 1500. The study is estimated to be completed in March 2021.

4.10 Clinical Evaluation of AZD3293

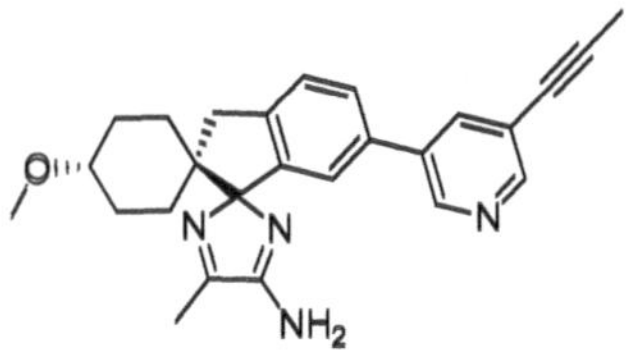

AZD3293

AstraZeneca brought this new compound to phase I clinical trials after their termination of AZD3839. AZD3293 is a member of the aminoimidazole class of nonpeptide inhibitors. A single-ascending dose trial of AZD3293 was completed in healthy male and female volunteers. AZD3293 was dosed ranging from 1 to 1000 mg and administered as an oral solution or placebo. Subjects were monitored over 10 days for adverse effects, variation in vital signs, and Columbia-Suicide Severity Rating Scale (C-SSRS). Plasma concentration of AZD3293 and biomarkers were also evaluated [136]. A two-part multiple-ascending dose study was performed on 47 subjects. Part one consisted of healthy elderly volunteers who were dosed with an oral solution of AZD3293 starting at 5 mg upward. Patients were monitored for adverse events, variation in vital signs and C-SSRS scores, as well as plasma and CSF concentrations of biomarkers and AZD3293. Part two consisted of mild-to-moderate AD patients demonstrating AD symptoms at least 6 months prior to the study. Subjects were dosed with an oral solution of AZD3293 and were examined for the same efficacy end points as part one [137].

Potential drug-drug interactions were investigated. AZD3293 was administered to healthy adult volunteers as an oral solution alongside itraconazole, diltiazem, and midazolam. Subjects were monitored for the effects of coadministration on pharmacokinetic parameters of AZD3293 [138]. Pharmacokinetic considerations were assessed in a study of healthy Japanese volunteers who received single doses as an oral solution, 15, 50, and 150 mg for young patients and 1, 15, and 50 mg for elderly patients. A multiple-dose cohort was also established. Subjects were monitored for adverse effects, pharmacokinetic concentrations of AZD3293, and $A\beta_{40}$ and $A\beta_{42}$ concentrations [139]. A study of the effect of AZD3293 on the QT interval was performed in 52 healthy male volunteers. A cohort of each high-dose and low-dose AZD3293 and a moxifloxacin and placebo cohort were given a single dose and monitored for changes in the QT interval as well as inhibitor plasma concentrations and incident of adverse events [140]. Additional phase I studies included a bioavailability assessment of the oral solution versus two different tablet formulations [141]; a study of the effect of food on these tablet formations [142]; an absorption, metabolism, and elimination study with ^{14}C-AZD3293 [143]; and drug-drug interaction studies with simvastatin, midazolam, and donepezil [144], warfarin [145], and dabigatran [146].

In September 2014, AstraZeneca announced it would collaborate with Eli Lilly to develop AZD3293 further and advance into phase II/III studies [147]. This study is currently recruiting participants with an estimated enrolment of 2,202 patients with mild AD. Participants will be given either a 20 or 50 mg dose of AZD3293 (LY3314814) orally, once daily for 104 weeks. Efficacy will be measured primarily by the change in CDR-SB score. Alternative measures include change in ADAS-Cog13 score, ADCS-ADL score, CDR score, NPI score, CSF $A\beta_{42}$ concentration, amyloid PET imaging, whole brain volume, and others [148]. This study is ongoing and results are expected in August 2021.

4.11 Clinical Evaluation of JNJ-54861911

In March 2013, JNJ-54861911 was brought into clinical development by Janssen Research and Development. Structural information for JNJ-54861911 has not been released. Healthy adult volunteers across eight cohorts totaling 56 subjects were given a single dose of JNJ-54861911 ranging from 1 to 160 mg or placebo, administered orally. Patient CSF and blood plasma were assessed for concentration of JNJ-54361911 and Aβ at up to 96 and 36 h post-dose, respectively [149]. A similar study was performed in healthy elderly volunteers. Seventy participants were given a 3–80 mg dose of JNJ-54861911 as an oral suspension, once daily for 14 days. Patients were monitored for adverse effects, while plasma and CSF samples were taken at multiple time points to assess concentration and exposure of JNJ-54861911 as well as Aβ levels [150]. To better understand the pharmacokinetic parameters of JNJ-54861911 and the effect food may have on these parameters, a study was performed in which healthy elderly male volunteers received a single 25 mg dose as an oral suspension under fasted conditions, a single 25 mg dose as an oral solid formulation under fasted conditions, and a single 25 mg dose as an oral solid formulation under fed conditions with 6 days between each dose and in varying orders of dosing. Bioavailability, absorption, half-life, elimination, and other pharmacokinetic parameters were examined using plasma concentration at discrete time points [151].

A proof-of-mechanism study was performed in early Alzheimer's patients. Forty-five participants were administered a 10 or 50 mg dose as an oral tablet once daily for 28 days. Exposure to JNJ-54861911 and biomarker levels in plasma and CSF were evaluated at discrete time points [152]. Other phase I studies included a study of pharmacokinetic considerations in healthy Japanese males [153] and at-risk asymptomatic Japanese participants [154]; drug-drug interaction studies with clarithromycin and itraconazole [155], caffeine and midazolam and tolbutamide [156], and rosuvastatin and metformin [157]; a study of the effects on QT intervals [158]; and the evaluation of the pharmacokinetic parameters of an alternative oral formulation [159].

Phase II trials for JNJ-54861911 began in November 2014. In a study of 114 participants with early AD, patients received a 10 mg, 50 mg, or placebo oral tablet once daily for 6 months. Safety and tolerability is the primary end point, with adverse events being monitored; however, exposure of JNJ-54861911 in the plasma and CSF as well as concentration of Aβ, sAPPβ, and sAPPα in the CSF and plasma will also be examined [160]. This study is expected to conclude in June 2016. A similar extended study began in July 2015. Patients received a once-daily, oral dose of either 10 mg or 25 mg for 96 weeks or longer until termination of study due to safety issue or registration of JNJ-54861911 [161]. The study is estimated to conclude in June 2024.

A phase II/III study began in October 2015 with an estimated 1,650 asymptomatic at-risk participants. Subjects will be administered 10 mg, 50 mg, or placebo once daily by mouth for 54 months. The primary outcome measurement will be the change in AD Cooperative Study-Preclinical Alzheimer's Cognitive Composite (ADCS-PACC) score, while secondary efficacy end points include change in CFI score and others, incident of adverse events, and concentration of JNJ-54861911 and biomarkers in the plasma and CSF [162]. This study is currently recruiting participants and is expected to be ongoing until May 2023.

4.12 Clinical Evaluation of VTP-37948

Vitae Pharmaceuticals discovered preclinical BACE1 inhibitor VTP-37948 (BI 1181181). This was licensed and developed clinically by Boehringer Ingelheim beginning in January 2014. The structure of VTP-37948 has not yet been disclosed. In phase I development, 65 healthy male volunteers were given a single dose of VTP-37948 in ascending cohorts. The study investigated safety and tolerability as well as pharmacokinetic parameters in three different formulations both with and without food by assessment of plasma and urine concentrations [163]. Another phase I study assessed pharmacokinetic parameters in healthy males via CSF and plasma analysis after single oral doses [164]. VTP-37948 was determined to be safe across all doses, with a half-life of 16–19 h; CSF Aβ levels could be lowered by 80% [165]. A multiple-ascending dose study was planned for healthy adult and healthy elderly volunteers. The trial began in September 2014, with participants receiving an oral dose once daily for 10 days [166]. In February 2015, it was announced that Vitae Pharmaceuticals and Boehringer Ingelheim were voluntarily terminating the clinical development of VTP-37948 due to negative skin reactions in some participants [167]. A drug-drug interaction study was withdrawn before participants could be recruited [168]. No further development or data for VTP-37948 has been reported.

4.13 Clinical Evaluation of CNP520

In partnership with Amgen, Novartis has examined CNP520 in phase II clinical trials as a single agent and in phase II/III clinical trials alongside immunotherapy CAD106. The structure of CNP520 has not yet been disclosed. The initial phase II trial was performed in healthy elderly volunteers. Four cohorts were administered a dose of CNP520 orally, once daily for 13 weeks. An additional cohort was given a placebo control. Patients were monitored for adverse effects as well as CNP520, $A\beta_{40}$, and $A\beta_{42}$ levels in the CSF and CNP520 concentrations in the plasma [169]. The study recently ended in March 2016. However, results have not yet been disclosed. A study of at-risk patients is currently recruiting participants to study CNP520 alongside immunotherapy CAD106 when administered separately. Participants will be given only CNP520 or CAD106 or a placebo, but not multiple investigational treatments. Volunteers given CNP520 will be dosed orally, once daily for 60 months. The study is expected to enroll 1340 participants and be completed in August 2023 [170].

5 Conclusions and Outlook

Since the cloning and characterization of BACE1 more than one and a half decades ago, BACE1 continues to be an exciting target for drug development against Alzheimer's disease. To date, several inhibitor drugs have now entered into clinical development for potential new treatment of Alzheimer's disease. This review highlighted design, development, and clinical evaluation of BACE1 inhibitors for the treatment of AD. The early design of BACE1 inhibitors involved replacing the scissile amide bond of the APP β-site with nonhydrolyzable transition-state isosteres. Subsequent determination of the X-ray crystal structure of these substrate-based inhibitors and BACE1 complexes provided necessary drug design templates and greatly facilitated the structure-based design of drug-like peptidomimetic BACE1 inhibitors. The structural evolution of BACE1 inhibitors has been very impressive. The availability of X-ray structural information of BACE1 with a variety of inhibitors along with past experiences of successful drug design against other aspartic acid proteases, such as renin and HIV protease, expedited the design and discovery of many new classes of BACE1 peptidomimetic inhibitors with therapeutic potential. Several BACE1 inhibitor drugs showed promising results in animal models, as they reduced the amyloid plaque in the brain and rescued cognitive deficits. The first peptidomimetic drug candidate, CTS21166, was evaluated in phase I human clinical trials. The reported results indicated that the drug was well tolerated, and Aβ reduction was achieved with BACE1 drug treatment.

Despite significant progress with peptidomimetic BACE1 inhibitors, AD drug development brought a different set of medicinal chemistry challenges. BACE1 has a large substrate-binding site which makes it difficult to design a small inhibitor that inhibits the enzyme with desirable drug-like properties. Compounds need to have sufficient lipophilic features to penetrate both plasma and intracellular membranes and gain access to the BACE1 active site in the brain. This requires inhibitor size to be around 550 Da with reduced susceptibility to Pgp-mediated efflux in order to cross the blood–brain barrier efficiently. Furthermore, the high homology among aspartyl proteases makes it particularly important to have high selectivity over cathepsin D, renin, and especially BACE2.

The X-ray structural knowledge of peptidomimetic inhibitors and extensive high-throughput screening efforts ultimately paved the way for the evolution of small-molecule nonpeptide BACE1 inhibitors. Many structural classes have now evolved with novel heterocyclic scaffolds such as acyl guanidine, aminopyridine, and aminoimidazole with unique modes of inhibition. Several optimized inhibitors from these classes have been shown to cross the BBB, reach the brain, and reduce brain Aβ in transgenic AD mice. As many as 13 such inhibitor drugs with favorable pharmacological properties have entered into clinical development, and two of these drugs have now advanced to phase IIa/b clinical trials with human AD patients. A successful clinical outcome would provide the long-waited proof for the "amyloid hypothesis" in AD therapy. The accumulation of knowledge of BACE1 and its functions through the years still supports BACE1 as a viable target for Alzheimer's disease drug development. The progress in the field allows much optimism that β-secretase inhibitor drug treatment of Alzheimer's disease may soon become a reality.

Acknowledgment The authors' work in the article was supported by the National Institutes of Health. We would also like to thank Ms. Anne Veitschegger and Mr. Luke Kassekert (both Purdue University) for their helpful discussions.

References

1. Selkoe DJ (2001) Alzheimer's disease: genes, proteins, and therapy. Physiol Rev 81(2): 741–766
2. Alzheimer's Association (2012) Alzheimer's disease facts and figures. Alzheimers Dement 8 (2):131–168. doi:10.1016/j.jalz.2012.02.001
3. Alzheimer's Association (2013) Alzheimer's disease facts and figures. Alzheimers Dement 9 (2):208–245. doi:10.1016/j.jalz.2013.02.003
4. Alzheimers.net (2016) 2016 Alzheimer's statistics. http://www.alzheimers.net/resources/alzheimers-statistics/. Accessed 20 May 2016
5. Cummings JL, Morstorf T, Zhong K (2014) Alzheimer's disease drug-development pipeline: few candidates, frequent failures. AlzheAlzheimers Res Ther 6(4):37. doi:10.1186/alzrt269
6. Anand P, Singh B (2013) A review on cholinesterase inhibitors for Alzheimer's disease. Arch Pharm Res 36(4):375–399. doi:10.1007/s12272-013-0036-3
7. Hansen RA, Gartlehner G, Webb AP, Morgan LC, Moore CG, Jonas DE (2008) Efficacy and safety of donepezil, galantamine, and rivastigmine for the treatment of Alzheimer's disease: a systematic review and meta-analysis. Clin Interv Aging 3(2):211–225
8. Olivares D, Deshpande VK, Shi Y, Lahiri DK, Greig NH, Rogers JT, Huang X (2012) N-methyl D-aspartate (NMDA) receptor antagonists and memantine treatment for Alzheimer's disease, vascular dementia and Parkinson's disease. Curr Alzheimer Res 9 (6):746–758. doi:10.2174/156720512801322564
9. Hyde C, Peters J, Bond M, Rogers G, Hoyle M, Anderson R, Jeffreys M, Davis S, Thokala P, Moxham T (2013) Evolution of the evidence on the effectiveness and cost-effectiveness of acetylcholinesterase inhibitors and memantine for Alzheimer's disease: systematic review and economic model. Age Ageing 42(1):14–20. doi:10.1093/ageing/afs165
10. Selkoe DJ (1999) Translating cell biology into therapeutic advances in Alzheimer's disease. Nature 399(6738):A23–A31. doi:10.1038/399a023
11. Ghosh AK, Osswald HL (2014) BACE1 (β-secretase) inhibitors for the treatment of Alzheimer's disease. Chem Soc Rev 43(19):6765–6813. doi:10.1039/c3cs60460h
12. Butini S, Brogi S, Novellino E, Campiani G, Ghosh AK, Brindisi M, Gemma S (2013) The structural evolution of β-secretase inhibitors: a focus on the development of small-molecule inhibitors. Curr Top Med Chem 13(15):1787–1807. doi:10.2174/15680266113139990137
13. Citron M (2010) Alzheimer's disease: strategies for disease modification. Nat Rev Drug Discov 9(5):387–398. doi:10.1038/nrd2896
14. Billings LM, Oddo S, Green KN, McGaugh JL, LaFerla FM (2005) Intraneuronal Aβ causes the onset of early Alzheimer's disease-related cognitive deficits in transgenic mice. Neuron 45(5):675–688. doi:10.1016/j.neuron.2005.01.040
15. Naslund J, Haroutunian V, Mohs R, Davis KL, Davies P, Greengard P, Buxbaum JD (2000) Correlation between elevated levels of amyloid-β peptide in the brain and cognitive decline. J Am Med Assoc 283(12):1571–1577. doi:10.1001/jama.283.12.1571
16. Ohno M, Cole SL, Yasvoina M, Zhao J, Citron M, Berry R, Disterhoft JF, Vassar R (2007) BACE1 gene deletion prevents neuron loss and memory deficits in 5XFAD APP/PS1 transgenic mice. Neurobiol Dis 26(1):134–145. doi:10.1016/j.nbd.2006.12.008
17. Wolfe MS (2008) Inhibition and modulation of γ-secretase for Alzheimer's disease. Neurotherapeutics 5(3):391–398. doi:10.1016/j.nurt.2008.05.010
18. Sinha S (2010) BACE: a (almost) perfect target for staving off Alzheimer's disease. In: Ghosh AK (ed) Aspartic acid proteases as therapeutic targets, vol 45, Methods and principles in medicinal chemistry. Wiley-VCH, Weinheim, pp 393–412
19. Tang J, Hong L, Ghosh AK (2010) The discovery of β-secretase and development toward a clinical inhibitor for AD: an exciting academic collaboration. In: Ghosh AK (ed) Aspartic acid proteases as therapeutic targets, vol 45, Methods and principles in medicinal chemistry. Wiley-VCH, Weinheim, pp 413–440

20. Selkoe DJ (1994) Cell biology of the amyloid β-protein precursor and the mechanism of Alzheimer's disease. Annu Rev Cell Biol 10:373–403. doi:10.1146/annurev.cb.10.110194.002105
21. Hardy J, Selkoe DJ (2002) The amyloid hypothesis of Alzheimer's disease: progress and problems on the road to therapeutics. Science 297:353–356. doi:10.1126/science.1072994
22. Selkoe DJ (2002) Alzheimer's disease is a synaptic failure. Science 298(5594):789–791. doi:10.1126/science.1074069
23. Selkoe DJ (2000) The origins of Alzheimer's disease: a is for amyloid. J Am Med Assoc 283 (12):1615–1617. doi:10.1001/jama.283.12.1615
24. Hussain I, Powell D, Howlett DR, Tew DG, Meek TD, Chapman C, Gloger IS, Murphy KE, Southan CD, Ryan DM, Smith TS, Simmons DL, Walsh FS, Dingwall C, Christie G (1999) Identification of a novel aspartic protease (Asp2) as β-secretase. Mol Cell Neurosci 14(6): 419–427. doi:10.1006/mcne.1999.0811
25. Sinha S, Anderson JP, Barbour R, Basi GS, Caccavello R, Davis D, Doan M, Dovey HF, Frigon N, Hong J, Jacobson-Croak K, Jewett N, Keim P, Knops J, Lieberburg I, Power M, Tan H, Tatsuno G, Tung J, Schenk D, Seubert P, Suomensaari SM, Wang S, Walker D, Zhao J, McConlogue L, John V (1999) Purification and cloning of amyloid precursor protein β-secretase from human brain. Nature 402(6761):537–540. doi:10.1038/990114
26. Lin X, Koelsch G, Wu S, Downs D, Dashti A, Tang J (2000) Human aspartic protease memapsin 2 cleaves the β-secretase site of β-amyloid precursor protein. Proc Natl Acad Sci U S A 97(4):1456–1460. doi:10.1073/pnas.97.4.1456
27. Vassar R, Bennett BD, Babu-Khan S, Kahn S, Mendiaz EA, Denis P, Teplow DB, Ross S, Amarante P, Loeloff R, Luo Y, Fisher S, Fuller J, Edenson S, Lile J, Jarosinski MA, Biere AL, Curran E, Burgess T, Louis JC, Collins F, Treanor J, Rogers G, Citron M (1999) β-secretase cleavage of Alzheimer's amyloid precursor protein by the transmembrane aspartic protease BACE. Science 286(5440):735–741. doi:10.1126/science.286.5440.735
28. Sinha S, Lieberburg I (1999) Cellular mechanisms of β-amyloid production and secretion. Proc Natl Acad Sci U S A 96(20):11049–11053. doi:10.1073/pnas.96.20.11049
29. Cai H, Wang Y, McCarthy D, Wen H, Borchelt DR, Price DL, Wong PC (2001) BACE1 is the major β-secretase for generation of Aβ peptides by neurons. Nat Neurosci 4(3):233–234. doi:10.1038/85064
30. Roberds SL, Anderson JP, Basi G, Bienkowski MJ, Branstetter DG, Chen KS, Freedman SB, Frigon NL, Games D, Hu K, Johnson-Wood K, Keappenman KE, Kawabe TT, Kola I, Kuehn R, Lee M, Liu W, Motter R, Nichols NF, Power M, Robertson DW, Schenk D, Schoor M, Shopp GM, Shuck ME, Sinha S, Svensson KA, Tatsuno G, Tintrup H, Wijsman J, Wright S, McConlogue L (2001) BACE Knockout mice are healthy despite lacking the primary β-secretase activity in brain: implications for Alzheimer's disease therapeutics. Hum Mol Genet 10(12):1317–1324. doi:10.1093/hmg/10.12.1317
31. Luo Y, Bolon B, Kahn S, Bennett BD, Babu-Khan S, Denis P, Fan W, Kha H, Zhang J, Gong Y, Martin L, Louis JC, Yan Q, Richards WG, Citron M, Vassar R (2001) Mice deficient in BACE1, the Alzheimer's β-secretase, have normal phenotype and abolished β-amyloid generation. Nat Neurosci 4(3):231–232. doi:10.1038/85059
32. Iserloh U, Cumming JN (2010) Peptidomimetic BACE1 inhibitors for treatment of Alzheimer's disease: design and evolution. In: Ghosh AK (ed) Aspartic acid proteases as therapeutic targets, vol 45, Methods and principles in medicinal chemistry. Wiley-VCH, Weinheim, pp 441–479
33. Nishitomi K, Sakaguchi G, Horikoshi Y, Gray AJ, Maeda M, Hirata-Fukae C, Becker AG, Hosono M, Sakaguchi I, Minami SS, Nakajima Y, Li HF, Takeyama C, Kihara T, Ota A, Wong PC, Aisen PS, Kato A, Kinoshita N, Matsuoka Y (2006) BACE1 inhibition reduces endogenous Aβ and alters APP processing in wild-type mice. J Neurochem 99(6):1555–1563. doi:10.1111/j.1471-4159.2006.04178.x

34. Hu X, Hicks CW, He W, Wong P, Macklin WB, Trapp BD, Yan R (2006) Bace1 modulates myelination in the central and peripheral nervous system. Nat Neurosci 9(12):1520–1525. doi:10.1038/nn1797
35. Hu X, Zhou X, He W, Yang J, Xiong W, Wong P, Wilson CG, Yan R (2010) BACE1 deficiency causes altered neuronal activity and neurodegeneration. J Neurosci 30 (26):8819–8829. doi:10.1523/JNEUROSCI.1334-10.2010
36. Kobayashi D, Zeller M, Cole T, Buttini M, McConlogue L, Sinha S, Freedman S, Morris RG, Chen KS (2008) BACE1 gene deletion: impact on behavioral function in a model of Alzheimer's disease. Neurobiol Aging 29(6):861–873. doi:10.1016/j.neurobiolaging.2007.01.002
37. Hitt B, Riordan SM, Kukreja L, Eimer WA, Rajapaksha TW, Vassar R (2012) beta-Site amyloid precursor protein (APP)-cleaving enzyme 1 (BACE1)-deficient mice exhibit a close homolog of L1 (CHL1) loss-of-function phenotype involving axon guidance defects. J Biol Chem 287(46):38408–38425. doi:10.1074/jbc.M112.415505
38. Laird FM, Cai H, Savonenko AV, Farah MH, He K, Melnikova T, Wen H, Chiang HC, Xu G, Koliatsos VE, Borchelt DR, Price DL, Lee HK, Wong PC (2005) BACE1, a major determinant of selective vulnerability of the brain to amyloid-beta amyloidogenesis, is essential for cognitive, emotional, and synaptic functions. J Neurosci 25(50):11693–11709
39. Savonenko AV, Melnikova T, Laird FM, Stewart KA, Price DL, Wong PC (2008) Alteration of BACE1-dependent NRG1/ErbB4 signaling and schizophrenia-like phenotypes in BACE1-null mice. Proc Natl Acad Sci U S A 105(14):5585–5590. doi:10.1073/pnas.0710373105
40. LaFerla FM, Green KN, Oddo S (2007) Intracellular amyloid-β in Alzheimer's disease. Nat Rev Neurosci 8(7):499–509
41. Chang WP, Koelsch G, Wong S, Downs D, Da H, Weerasena V, Gordon B, Devasamudram T, Bilcer G, Ghosh AK, Tang J (2004) In vivo inhibition of Aβ production by memapsin 2 (β-secretase) inhibitors. J Neurochem 89(6):1409–1416. doi:10.1111/j.1471-4159.2004.02452.x
42. Rajendran L, Schneider A, Schlechtingen G, Weidlich S, Ries J, Braxmeier T, Schwille P, Schulz JB, Schroeder C, Simons M, Jennings G, Knolker HJ, Simons K (2008) Efficient inhibition of the Alzheimer's disease β-secretase by membrane targeting. Science 320(5875): 520–523. doi:10.1126/science.1156609
43. Meredith JE Jr, Thompson LA, Toyn JH, Marcin L, Barten DM, Marcinkeviciene J, Kopcho L, Kim Y, Lin A, Guss V, Burton C, Iben L, Polson C, Cantone J, Ford M, Drexler D, Fiedler T, Lentz KA, Grace JE Jr, Kolb J, Corsa J, Pierdomenico M, Jones K, Olson RE, Macor JE, Albright CF (2008) P-glycoprotein efflux and other factors limit brain amyloid-β reduction by β-site amyloid precursor protein-cleaving enzyme 1 inhibitors in mice. J Pharmacol Exp Ther 326(2):502–513. doi:10.1124/jpet.108.138974
44. Hong L, Tang J (2004) Flap position of free memapsin 2 (β-secretase), a model for flap opening in aspartic protease catalysis. Biochemistry 43(16):4689–4695. doi:10.1021/bi0498252
45. Hong L, Koelsch G, Lin X, Wu S, Terzyan S, Ghosh AK, Zhang XC, Tang J (2000) Structure of the protease domain of memapsin 2 (β-secretase) complexed with inhibitor. Science 290 (5489):150–153. doi:10.1126/science.290.5489.150
46. Hong L, Turner RT 3rd, Koelsch G, Shin D, Ghosh AK, Tang J (2002) Crystal structure of memapsin 2 (β-secretase) in complex with an inhibitor OM00-3. Biochemistry 41(36): 10963–10967. doi:10.1021/bi026232n
47. Bennett BD, Babu-Khan S, Loeloff R, Louis JC, Curran E, Citron M, Vassar R (2000) Expression analysis of BACE2 in brain and peripheral tissues. J Biol Chem 275(27): 20647–20651. doi:10.1074/jbc.M002688200
48. Ghosh AK, Brindisi M, Tang J (2012) Developing β-secretase inhibitors for treatment of Alzheimer's disease. J Neurochem 120(Suppl 1):71–83. doi:10.1111/j.1471-4159.2011.07476.x

49. Turner RT 3rd, Koelsch G, Hong L, Castanheira P, Ermolieff J, Ghosh AK, Tang J (2001) Subsite specificity of memapsin 2 (β-secretase): implications for inhibitor design. Biochemistry 40(34):10001–10006. doi:10.1021/bi015546s
50. Ghosh AK, Shin DW, Downs D, Koelsch G, Lin XL, Ermolieff J, Tang J (2000) Design of potent inhibitors for human brain memapsin 2 (β-secretase). J Am Chem Soc 122(14): 3522–3523. doi:10.1021/Ja000300g
51. Ghosh AK, Bilcer G, Harwood C, Kawahama R, Shin D, Hussain KA, Hong L, Loy JA, Nguyen C, Koelsch G, Ermolieff J, Tang J (2001) Structure-based design: potent inhibitors of human brain memapsin 2 (β-secretase). J Med Chem 44(18):2865–2868. doi:10.1021/jm0101803
52. Ghosh AK, Kumaragurubaran N, Hong L, Lei H, Hussain KA, Liu CF, Devasamudram T, Weerasena V, Turner R, Koelsch G, Bilcer G, Tang J (2006) Design, synthesis and X-ray structure of protein-ligand complexes: important insight into selectivity of memapsin 2 (-β-secretase) inhibitors. J Am Chem Soc 128(16):5310–5311. doi:10.1021/ja058636j
53. Ghosh AK, Kumaragurubaran N, Hong L, Kulkarni SS, Xu XM, Chang WP, Weerasena V, Turner R, Koelsch G, Bilcer G, Tang J (2007) Design, synthesis, and X-ray structure of potent memapsin 2 (β-secretase) inhibitors with isophthalamide derivatives as the P2-P3-ligands. J Med Chem 50(10):2399–2407. doi:10.1021/Jm061338s
54. Bjorklund C, Oscarson S, Benkestock K, Borkakoti N, Jansson K, Lindberg J, Vrang L, Hallberg A, Rosenquist A, Samuelsson B (2010) Design and synthesis of potent and selective BACE-1 inhibitors. J Med Chem 53(4):1458–1464. doi:10.1021/jm901168f
55. Ghosh AK, Chapsal BD, Mitsuya H (2010) Darunavir, a new PI with dual mechanism: from a novel drug design concept to new hope against drug-resistant HIV. In: Ghosh AK (ed) Aspartic acid proteases as therapeutic targets, vol 45, Methods and principles in medicinal chemistry. Wiley-VCH, Weinheim, pp 204–243. doi:10.1002/9783527630943.ch8
56. Ghosh AK, Anderson DD, Mitsuya H (2010) The FDA approved HIV-1 protease inhibitors for treatment of HIV/AIDS. In: Burger's medicinal chemistry and drug discovery. John Wiley & Sons, Inc., pp 1–74. doi:10.1002/0471266949.bmc224
57. Tamamura H, Kato T, Otaka A, Fujii N (2003) Synthesis of potent β-secretase inhibitors containing a hydroxyethylamine dipeptide isostere and their structure-activity relationship studies. Org Biomol Chem 1(14):2468–2473. doi:10.1039/b304842j
58. Stachel SJ, Coburn CA, Steele TG, Jones KG, Loutzenhiser EF, Gregro AR, Rajapakse HA, Lai MT, Crouthamel MC, Xu M, Tugusheva K, Lineberger JE, Pietrak BL, Espeseth AS, Shi XP, Chen-Dodson E, Holloway MK, Munshi S, Simon AJ, Kuo L, Vacca JP (2004) Structure-based design of potent and selective cell-permeable inhibitors of human β-secretase (BACE-1). J Med Chem 47(26):6447–6450. doi:10.1021/jm049379g
59. Chang WP, Huang X, Downs D, Cirrito JR, Koelsch G, Holtzman DM, Ghosh AK, Tang J (2011) β-secretase inhibitor GRL-8234 rescues age-related cognitive decline in APP transgenic mice. FASEB J 25(2):775–784. doi:10.1096/fj.10-167213
60. Ghosh AK, Kumaragurubaran N, Hong L, Kulkarni S, Xu X, Miller HB, Reddy DS, Weerasena V, Turner R, Chang W, Koelsch G, Tang J (2008) Potent memapsin 2 (β-secretase) inhibitors: design, synthesis, protein-ligand X-ray structure, and in vivo evaluation. Bioorg Med Chem Lett 18(3):1031–1036. doi:10.1016/j.bmcl.2007.12.028
61. Beswick P, Charrier N, Clarke B, Demont E, Dingwall C, Dunsdon R, Faller A, Gleave R, Hawkins J, Hussain I, Johnson CN, MacPherson D, Maile G, Matico R, Milner P, Mosley J, Naylor A, O'Brien A, Redshaw S, Riddell D, Rowland P, Skidmore J, Soleil V, Smith KJ, Stanway S, Stemp G, Stuart A, Sweitzer S, Theobald P, Vesey D, Walter DS, Ward J, Wayne G (2008) BACE-1 inhibitors part 3: identification of hydroxy ethylamines (HEAs) with nanomolar potency in cells. Bioorg Med Chem Lett 18(3):1022–1026. doi:10.1016/j.bmcl.2007.12.020
62. Charrier N, Clarke B, Cutler L, Demont E, Dingwall C, Dunsdon R, Hawkins J, Howes C, Hubbard J, Hussain I, Maile G, Matico R, Mosley J, Naylor A, O'Brien A, Redshaw S, Rowland P, Soleil V, Smith KJ, Sweitzer S, Theobald P, Vesey D, Walter DS, Wayne G

(2009) Second generation of BACE-1 inhibitors part 3: towards non hydroxyethylamine transition state mimetics. Bioorg Med Chem Lett 19(13):3674–3678. doi:10.1016/j.bmcl.2009.03.149

63. Rajapakse HA, Nantermet PG, Selnick HG, Munshi S, McGaughey GB, Lindsley SR, Young MB, Lai MT, Espeseth AS, Shi XP, Colussi D, Pietrak B, Crouthamel MC, Tugusheva K, Huang Q, Xu M, Simon AJ, Kuo L, Hazuda DJ, Graham S, Vacca JP (2006) Discovery of oxadiazoyl tertiary carbinamine inhibitors of β-secretase (BACE-1). J Med Chem 49(25): 7270–7273. doi:10.1021/jm061046r
64. Sankaranarayanan S, Holahan MA, Colussi D, Crouthamel MC, Devanarayan V, Ellis J, Espeseth A, Gates AT, Graham SL, Gregro AR, Hazuda D, Hochman JH, Holloway K, Jin L, Kahana J, Lai MT, Lineberger J, McGaughey G, Moore KP, Nantermet P, Pietrak B, Price EA, Rajapakse H, Stauffer S, Steinbeiser MA, Seabrook G, Selnick HG, Shi XP, Stanton MG, Swestock J, Tugusheva K, Tyler KX, Vacca JP, Wong J, Wu G, Xu M, Cook JJ, Simon AJ (2009) First demonstration of cerebrospinal fluid and plasma Aβ lowering with oral administration of a β-site amyloid precursor protein-cleaving enzyme 1 inhibitor in nonhuman primates. J Pharmacol Exp Ther 328(1):131–140. doi:10.1124/jpet.108.143628
65. Nantermet PG, Rajapakse HA, Stanton MG, Stauffer SR, Barrow JC, Gregro AR, Moore KP, Steinbeiser MA, Swestock J, Selnick HG, Graham SL, McGaughey GB, Colussi D, Lai MT, Sankaranarayanan S, Simon AJ, Munshi S, Cook JJ, Holahan MA, Michener MS, Vacca JP (2009) Evolution of tertiary carbinamine BACE-1 inhibitors: Aβ reduction in rhesus CSF upon oral dosing. ChemMedChem 4(1):37–40. doi:10.1002/cmdc.200800308
66. Stanton MG, Stauffer SR, Gregro AR, Steinbeiser M, Nantermet P, Sankaranarayanan S, Price EA, Wu G, Crouthamel MC, Ellis J, Lai MT, Espeseth AS, Shi XP, Jin L, Colussi D, Pietrak B, Huang Q, Xu M, Simon AJ, Graham SL, Vacca JP, Selnick H (2007) Discovery of isonicotinamide derived β-secretase inhibitors: in vivo reduction of β-amyloid. J Med Chem 50(15):3431–3433. doi:10.1021/jm070272d
67. Coburn CA, Stachel SJ, Jones KG, Steele TG, Rush DM, DiMuzio J, Pietrak BL, Lai MT, Huang Q, Lineberger J, Jin L, Munshi S, Katharine Holloway M, Espeseth A, Simon A, Hazuda D, Graham SL, Vacca JP (2006) BACE-1 inhibition by a series of ψ[CH_2NH] reduced amide isosteres. Bioorg Med Chem Lett 16(14):3635–3638. doi:10.1016/j.bmcl.2006.04.076
68. Ghosh AK, Venkateswara Rao K, Yadav ND, Anderson DD, Gavande N, Huang X, Terzyan S, Tang J (2012) Structure-based design of highly selective β-secretase inhibitors: synthesis, biological evaluation, and protein-ligand X-ray crystal structure. J Med Chem 55 (21): 9195–9207. doi:10.1021/jm3008823
69. Ghosh AK, Devasamudram T, Hong L, DeZutter C, Xu X, Weerasena V, Koelsch G, Bilcer G, Tang J (2005) Structure-based design of cycloamide-urethane-derived novel inhibitors of human brain memapsin 2 (β-secretase). Bioorg Med Chem Lett 15(1):15–20. doi:10.1016/j.bmcl.2004.10.084
70. Machauer R, Laumen K, Veenstra S, Rondeau JM, Tintelnot-Blomley M, Betschart C, Jaton AL, Desrayaud S, Staufenbiel M, Rabe S, Paganetti P, Neumann U (2009) Macrocyclic peptidomimetic β-secretase (BACE-1) inhibitors with activity in vivo. Bioorg Med Chem Lett 19(5): 1366–1370. doi:10.1016/j.bmcl.2009.01.055
71. Stachel SJ, Coburn CA, Sankaranarayanan S, Price EA, Wu G, Crouthamel M, Pietrak BL, Huang Q, Lineberger J, Espeseth AS, Jin L, Ellis J, Holloway MK, Munshi S, Allison T, Hazuda D, Simon AJ, Graham SL, Vacca JP (2006) Macrocyclic inhibitors of β-secretase: functional activity in an animal model. J Med Chem 49(21):6147–6150. doi:10.1021/jm060884i
72. Cole DC, Bursavich MG (2010) Nonpeptide BACE1 inhibitors: design and synthesis. In: Ghosh AK (ed) Aspartic acid proteases as therapeutic targets, vol 45, Methods and principles in medicinal chemistry. Wiley-VCH, Weinheim, pp 481–509
73. Cole DC, Manas ES, Stock JR, Condon JS, Jennings LD, Aulabaugh A, Chopra R, Cowling R, Ellingboe JW, Fan KY, Harrison BL, Hu Y, Jacobsen S, Jin G, Lin L, Lovering FE, Malamas

MS, Stahl ML, Strand J, Sukhdeo MN, Svenson K, Turner MJ, Wagner E, Wu J, Zhou P, Bard J (2006) Acylguanidines as small-molecule β-secretase inhibitors. J Med Chem 49 (21):6158–6161. doi:10.1021/jm0607451

74. Gerritz SW, Zhai W, Shi S, Zhu S, Toyn JH, Meredith JE Jr, Iben LG, Burton CR, Albright CF, Good AC, Tebben AJ, Muckelbauer JK, Camac DM, Metzler W, Cook LS, Padmanabha R, Lentz KA, Sofia MJ, Poss MA, Macor JE, Thompson LA 3rd (2012) Acyl guanidine inhibitors of β-secretase (BACE-1): optimization of a micromolar hit to a nanomolar lead via iterative solid- and solution-phase library synthesis. J Med Chem 55 (21):9208–9223. doi:10.1021/jm300931y
75. Thomas AA, Hunt KW, Volgraf M, Watts RJ, Liu X, Vigers G, Smith D, Sammond D, Tang TP, Rhodes SP, Metcalf AT, Brown KD, Otten JN, Burkard M, Cox AA, Do MKG, Dutcher D, Rana S, DeLisle RK, Regal K, Wright AD, Groneberg R, Scearce-Levie K, Siu M, Purkey HE, Lyssikatos JP, Gunawardana IW (2014) Discovery of 7-tetrahydropyran-2-yl chromans: β-site amyloid precursor protein cleaving enzyme 1 (BACE1) inhibitors that reduce amyloid β-protein (Aβ) in the central nervous system. J Med Chem 57(3):878–902. doi:10.1021/jm401635n
76. Congreve M, Aharony D, Albert J, Callaghan O, Campbell J, Carr RA, Chessari G, Cowan S, Edwards PD, Frederickson M, McMenamin R, Murray CW, Patel S, Wallis N (2007) Application of fragment screening by X-ray crystallography to the discovery of aminopyridines as inhibitors of β-secretase. J Med Chem 50(6):1124–1132. doi:10.1021/jm061197u
77. Malamas MS, Barnes K, Hui Y, Johnson M, Lovering F, Condon J, Fobare W, Solvibile W, Turner J, Hu Y, Manas ES, Fan K, Olland A, Chopra R, Bard J, Pangalos MN, Reinhart P, Robichaud AJ (2010) Novel pyrrolyl 2-aminopyridines as potent and selective human β-secretase (BACE1) inhibitors. Bioorg Med Chem Lett 20(7):2068–2073. doi:10.1016/j.bmcl.2010.02.075
78. Hills ID, Holloway MK, de Leon P, Nomland A, Zhu H, Rajapakse H, Allison TJ, Munshi SK, Colussi D, Pietrak BL, Toolan D, Haugabook SJ, Graham SL, Stachel SJ (2009) A conformational constraint improves a β-secretase inhibitor but for an unexpected reason. Bioorg Med Chem Lett 19(17):4993–4995. doi:10.1016/j.bmcl.2009.07.071
79. Malamas MS, Erdei J, Gunawan I, Barnes K, Johnson M, Hui Y, Turner J, Hu Y, Wagner E, Fan K, Olland A, Bard J, Robichaud AJ (2009) Aminoimidazoles as potent and selective human β-secretase (BACE1) inhibitors. J Med Chem 52(20):6314–6323. doi:10.1021/jm9006752
80. Swahn BM, Holenz J, Kihlstrom J, Kolmodin K, Lindstrom J, Plobeck N, Rotticci D, Sehgelmeble F, Sundstrom M, Berg S, Falting J, Georgievska B, Gustavsson S, Neelissen J, Ek M, Olsson LL, Berg S (2012) Aminoimidazoles as BACE-1 inhibitors: the challenge to achieve in vivo brain efficacy. Bioorg Med Chem Lett 22(5):1854–1859. doi:10.1016/j.bmcl.2012.01.079
81. Malamas MS, Erdei J, Gunawan I, Turner J, Hu Y, Wagner E, Fan K, Chopra R, Olland A, Bard J, Jacobsen S, Magolda RL, Pangalos M, Robichaud AJ (2010) Design and synthesis of 5,5′-disubstituted aminohydantoins as potent and selective human β-secretase (BACE1) inhibitors. J Med Chem 53(3):1146–1158. doi:10.1021/jm901414e
82. Caldwell JP, Mazzola RD, Durkin J, Chen J, Chen X, Favreau L, Kennedy M, Kuvelkar R, Lee J, McHugh N, McKittrick B, Orth P, Stamford A, Strickland C, Voigt J, Wang L, Zhang L, Zhang Q, Zhu Z (2014) Discovery of potent iminoheterocycle BACE1 inhibitors. Bioorg Med Chem Lett 24(23):5455–5459. doi:10.1016/j.bmcl.2014.10.006
83. Egbertson M, McGaughey GB, Pitzenberger SM, Stauffer SR, Coburn CA, Stachel SJ, Yang W, Barrow JC, Neilson LA, McWherter M, Perlow D, Fahr B, Munshi S, Allison TJ, Holloway K, Selnick HG, Yang Z, Swestock J, Simon AJ, Sankaranarayanan S, Colussi D, Tugusheva K, Lai M-T, Pietrak B, Haugabook S, Jin L, Chen IW, Holahan M, Stranieri-Michener M, Cook JJ, Vacca J, Graham SL (2015) Methyl-substitution of an iminohydantoin

spiropiperidine β-secretase (BACE-1) inhibitor has a profound effect on its potency. Bioorg Med Chem Lett 25(21):4812–4819. doi:10.1016/j.bmcl.2015.06.082
84. Woltering TJ, Wostl W, Hilpert H, Rogers-Evans M, Pinard E, Mayweg A, Gobel M, Banner DW, Benz J, Travagli M, Pollastrini M, Marconi G, Gabellieri E, Guba W, Mauser H, Andreini M, Jacobsen H, Power E, Narquizian R (2013) BACE1 inhibitors: a head group scan on a series of amides. Bioorg Med Chem Lett 23(14):4239–4243. doi:10.1016/j.bmcl.2013.05.003
85. Hilpert H, Guba W, Woltering TJ, Wostl W, Pinard E, Mauser H, Mayweg AV, Rogers-Evans M, Humm R, Krummenacher D, Muser T, Schnider C, Jacobsen H, Ozmen L, Bergadano A, Banner DW, Hochstrasser R, Kuglstatter A, David-Pierson P, Fischer H, Polara A, Narquizian R (2013) β-Secretase (BACE1) inhibitors with high in vivo efficacy suitable for clinical evaluation in Alzheimer's disease. J Med Chem 56(10):3980–3995. doi:10.1021/jm400225m
86. Epstein O, Bryan MC, Cheng AC, Derakhchan K, Dineen TA, Hickman D, Hua Z, Human JB, Kreiman C, Marx IE, Weiss MM, Wahl RC, Wen PH, Whittington DA, Wood S, Zheng XM, Fremeau RT, White RD, Patel VF (2014) Lead optimization and modulation of hERG activity in a series of aminooxazoline xanthene β-site amyloid precursor protein cleaving enzyme (BACE1) inhibitors. J Med Chem 57(23):9796–9810. doi:10.1021/jm501266w
87. Cheng Y, Brown J, Judd TC, Lopez P, Qian W, Powers TS, Chen JJ, Bartberger MD, Chen K, Dunn RT, Epstein O, Fremeau RT, Harried S, Hickman D, Hitchcock SA, Luo Y, Minatti AE, Patel VF, Vargas HM, Wahl RC, Weiss MM, Wen PH, White RD, Whittington DA, Zheng XM, Wood S (2015) An orally available BACE1 inhibitor that affords robust CNS Aβ reduction without cardiovascular liabilities. ACS Med Chem Lett 6(2):210–215. doi:10.1021/ml500458t
88. Baxter EW, Conway KA, Kennis L, Bischoff F, Mercken MH, Winter HL, Reynolds CH, Tounge BA, Luo C, Scott MK, Huang Y, Braeken M, Pieters SM, Berthelot DJ, Masure S, Bruinzeel WD, Jordan AD, Parker MH, Boyd RE, Qu J, Alexander RS, Brenneman DE, Reitz AB (2007) 2-Amino-3,4-dihydroquinazolines as inhibitors of BACE-1 (beta-site APP cleaving enzyme): use of structure based design to convert a micromolar hit into a nanomolar lead. J Med Chem 50(18):4261–4264. doi:10.1021/jm0705408
89. Ghosh AK, Pandey S, Gangarajula S, Kulkarni S, Xu X, Rao KV, Huang X, Tang J (2012) Structure-based design, synthesis, and biological evaluation of dihydroquinazoline-derived potent β-secretase inhibitors. Bioorg Med Chem Lett 22(17):5460–5465. doi:10.1016/j.bmcl.2012.07.043
90. Cheng Y, Judd TC, Bartberger MD, Brown J, Chen K, Fremeau RT Jr, Hickman D, Hitchcock SA, Jordan B, Li V, Lopez P, Louie SW, Luo Y, Michelsen K, Nixey T, Powers TS, Rattan C, Sickmier EA, St Jean DJ Jr, Wahl RC, Wen PH, Wood S (2011) From fragment screening to in vivo efficacy: optimization of a series of 2-aminoquinolines as potent inhibitors of beta-site amyloid precursor protein cleaving enzyme 1 (BACE1). J Med Chem 54(16):5836–5857. doi:10.1021/jm200544q
91. Stachel SJ, Steele TG, Petrocchi A, Haugabook SJ, McGaughey G, Katharine Holloway M, Allison T, Munshi S, Zuck P, Colussi D, Tugasheva K, Wolfe A, Graham SL, Vacca JP (2012) Discovery of pyrrolidine-based β-secretase inhibitors: lead advancement through conformational design for maintenance of ligand binding efficiency. Bioorg Med Chem Lett 22(1): 240–244. doi:10.1016/j.bmcl.2011.11.024
92. Huang Y, Strobel ED, Ho CY, Reynolds CH, Conway KA, Piesvaux JA, Brenneman DE, Yohrling GJ, Moore Arnold H, Rosenthal D, Alexander RS, Tounge BA, Mercken M, Vandermeeren M, Parker MH, Reitz AB, Baxter EW (2010) Macrocyclic BACE inhibitors: optimization of a micromolar hit to nanomolar leads. Bioorg Med Chem Lett 20(10): 3158–3160. doi:10.1016/j.bmcl.2010.03.097
93. Boy KM, Guernon JM, Wu Y-J, Zhang Y, Shi J, Zhai W, Zhu S, Gerritz SW, Toyn JH, Meredith JE, Barten DM, Burton CR, Albright CF, Good AC, Grace JE, Lentz KA, Olson RE, Macor JE, Thompson Iii LA (2015) Macrocyclic prolinyl acyl guanidines as inhibitors of

β-secretase (BACE). Bioorg Med Chem Lett 25(22):5040–5047. doi:10.1016/j.bmcl.2015.10.031

94. CoMentis. Safety study of CTS21166 to treat Alzheimer's disease. In: ClinicalTrials.gov [Internet]. Bethesda (MD): National Library of Medicine (US). 2000- [2016 Apr 28]. Available from: https://clinicaltrials.gov/show/NCT00621010
95. Koelsch G (2008) Beta-secretase inhibitor CTS-21166 reduces plasma $A\beta_{40}$ in human subjects. In: Keystone symposium on Alzheimer's disease, Keystone, Colorado, USA, 2008
96. Strobel G (2008) Keystone drug news: CoMentis BACE inhibitor debuts. Alzforum. http://www.alzforum.org/news/conference-coverage/keystone-drug-news-comentis-bace-inhibitor-debuts?id=1790. Accessed 25 July 2016
97. Eli Lilly and Company. A safety study of LY2811376 single doses in healthy subjects. In: ClinicalTrials.gov [Internet]. Bethesda (MD): National Library of Medicine (US). 2000- [2016 Apr 28]. Available from: https://clinicaltrials.gov/show/NCT00838084
98. May PC, Dean RA, Lowe SL, Martenyi F, Sheehan SM, Boggs LN, Monk SA, Mathes BM, Mergott DJ, Watson BM, Stout SL, Timm DE, Smith Labell E, Gonzales CR, Nakano M, Jhee SS, Yen M, Ereshefsky L, Lindstrom TD, Calligaro DO, Cocke PJ, Greg Hall D, Friedrich S, Citron M, Audia JE (2011) Robust central reduction of amyloid-β in humans with an orally available, non-peptidic β-secretase inhibitor. J Neurosci 31(46):16507–16516. doi:10.1523/JNEUROSCI.3647-11.2011
99. LY2886721. Alzforum. Available via Biomedical Research Forum, LLC. http://www.alzforum.org/therapeutics/ly2886721. Accessed 29 Apr 2016
100. Eli Lilly and Company. A safety study of LY2886721 single doses in healthy subjects. In: ClinicalTrials.gov [Internet]. Bethesda (MD): National Library of Medicine (US). 2000- [2016 Apr 29]. Available from: https://clinicaltrials.gov/show/NCT01133405
101. Eli Lilly and Company. Multiple-ascending dose, safety, tolerability, pharmacokinetic, and pharmacodynamic study of LY2886721 in healthy subjects. In: ClinicalTrials.gov [Internet]. Bethesda (MD): National Library of Medicine (US). 2000- [2016 Apr 29]. Available from: https://clinicaltrials.gov/show/NCT01227252
102. Eli Lilly and Company. Disposition of ^{14}C-LY2886721 following oral administration in healthy human subjects. In: ClinicalTrials.gov [Internet]. Bethesda (MD): National Library of Medicine (US). 2000- [2016 Apr 29]. Available from: https://clinicaltrials.gov/show/NCT01367262
103. Eli Lilly and Company. Single- and multiple-dose, safety, tolerability, pharmacokinetic, and pharmacodynamic study of LY2886721 in healthy subjects. In: ClinicalTrials.gov [Internet]. Bethesda (MD): National Library of Medicine (US). 2000- [2016 Apr 29]. Available from: https://clinicaltrials.gov/show/NCT01534273
104. Eli Lilly and Company. Assessment of safety, tolerability, and pharmacodynamic effects of LY2886721 in patients with mild cognitive impairment due to Alzheimer's disease or mild Alzheimer's disease. In: ClinicalTrials.gov [Internet]. Bethesda (MD): National Library of Medicine (US). 2000- [2016 Apr 29]. Available from: https://clinicaltrials.gov/show/NCT01561430
105. Eli Lilly and Company. A comparison study of capsule and orally disintegrating tablet and to determine the effect of food and water on the pharmacokinetics of LY2886721 in healthy subjects. In: ClinicalTrials.gov [Internet]. Bethesda (MD): National Library of Medicine (US). 2000- [2016 Apr 29]. Available from: https://clinicaltrials.gov/show/NCT01775904
106. Eli Lilly and Company (2013) Lilly voluntarily terminates phase II study for LY2886721, a β-secretase inhibitor, being investigated as a treatment for Alzheimer's disease. https://investor.lilly.com/releasedetail.cfm?releaseid=771353
107. Landhuis E (2012) Wave of new BACE inhibitors heading to phase 2. Biomedical Research Forum, LLC. Available via Alzforum. http://www.alzforum.org/news/conference-coverage/wave-new-bace-inhibitors-heading-phase-2. Accessed 29 Apr 2016
108. Eisai Inc. A randomized, double-blind, placebo-controlled, single ascending dose study to evaluate the safety, tolerability, pharmacokinetics, and pharmacodynamics of oral doses of

E2609 in healthy subjects and an elderly cohort. In: ClinicalTrials.gov [Internet]. Bethesda (MD): National Library of Medicine (US). 2000- [2016 Apr 29]. Available from: https://clinicaltrials.gov/show/NCT01294540
109. Lai R, Albala B, Kaplow JM, Aluri J, Yen M, Satlin A. First-in-human study of E2609, a novel BACE1 inhibitor, demonstrates prolonged reductions in plasma β-amyloid levels after single dosing. Alzheimers Dement 8(4):P96. doi:10.1016/j.jalz.2012.05.237
110. Eisai Inc. A randomized, double-blind, placebo-controlled, multiple ascending dose study to evaluate the safety, tolerability, pharmacokinetics, and pharmacodynamics of E2609 in healthy subjects. In: ClinicalTrials.gov [Internet]. Bethesda (MD): National Library of Medicine (US). 2000- [2016 Apr 29]. Available from: https://clinicaltrials.gov/show/NCT01511783
111. Albala B, Kaplow JM, Lai R, Matijevic M, Aluri J, Satlin A. CSF amyloid lowering in human volunteers after 14 days' oral administration of the novel BACE1 inhibitor E2609. Alzheimers Dement 8(4):S743. doi:10.1016/j.jalz.2013.08.023
112. Eisai Inc. A randomized, double-blind, placebo-controlled study to evaluate the safety, tolerability, pharmacokinetics, and pharmacodynamics of single oral doses of E2609 in subjects with mild cognitive impairment or mild dementia due to Alzheimer's disease. In: ClinicalTrials.gov [Internet]. Bethesda (MD): National Library of Medicine (US). 2000- [2016 Apr 29]. Available from: https://clinicaltrials.gov/show/NCT01600859
113. Eisai Inc. A randomized, open-label, 3-treatment crossover study to determine the bioavailability of E2609 tablets compared to capsules and the effect of food on absorption in healthy Caucasian male adults. In: ClinicalTrials.gov [Internet]. Bethesda (MD): National Library of Medicine (US). 2000- [2016 Apr 29]. Available from: https://clinicaltrials.gov/show/NCT01716897
114. Eisai Inc. An open-label, single dose study to determine the metabolism and elimination or [^{14}C]E2609 in healthy male subjects. In: ClinicalTrials.gov [Internet]. Bethesda (MD): National Library of Medicine (US). 2000- [2016 Apr 29]. Available from: https://clinicaltrials.gov/show/NCT01975636
115. Eisai Inc. A 3-part, open-label, drug-drug interaction study of concomitant administration of E2609 with itraconazole, rifampin, digoxin, or donepezil. In: ClinicalTrials.gov [Internet]. Bethesda (MD): National Library of Medicine (US). 2000- [2016 Apr 29]. Available from: https://clinicaltrials.gov/show/NCT02055703
116. Eisai Inc. Study to evaluate the safety, tolerability, pharmacokinetics, and pharmacodynamics of E2609 in healthy adult male Japanese and white subjects. In: ClinicalTrials.gov [Internet]. Bethesda (MD): National Library of Medicine (US). 2000- [2016 Apr 29]. Available from: https://clinicaltrials.gov/show/NCT02207790
117. Eisai Inc. A randomized, double-blind, placebo and active-controlled, single-dose, 4-treatment crossover study to evaluate the effects of E2609 on QTc interval in healthy subjects. In: ClinicalTrials.gov [Internet]. Bethesda (MD): National Library of Medicine (US). 2000- [2016 Apr 29]. Available from: https://clinicaltrials.gov/show/NCT02222324
118. Eisai Inc, Biogen. A placebo-controlled, double-blind, parallel-group, randomized, proof-of-concept, dose-finding study to evaluate safety, tolerability, and efficacy of E2609 in subjects with mild cognitive impairment due to Alzheimer's disease (prodromal Alzheimer's disease) and mild to moderate dementia due to Alzheimer's disease. In: ClinicalTrials.gov [Internet]. Bethesda (MD): National Library of Medicine (US). 2000- [2016 Apr 29]. Available from: https://clinicaltrials.gov/show/NCT02322021
119. Jeppsson F, Eketjall S, Janson J, Karlstrom S, Gustavsson S, Olsson LL, Radesater AC, Ploeger B, Cebers G, Kolmodin K, Swahn BM, von Berg S, Bueters T, Falting J (2012) Discovery of AZD3839, a potent and selective BACE1 inhibitor clinical candidate for the treatment of Alzheimer's disease. J Biol Chem 287(49):41245–41257. doi:10.1074/jbc.M112.409110
120. AstraZeneca. A phase I, randomized, double-blind, placebo-controlled, parallel-group, single ascending dose study to assess the safety, tolerability, pharmacokinetics, and effect on

biomarkers of AZD3839 including an open-label food effect group in healthy male and female volunteers of non-childbearing potential. In: ClinicalTrials.gov [Internet]. Bethesda (MD): National Library of Medicine (US). 2000- [2016 May 02]. Available from: https://clinicaltrials.gov/show/NCT01348737

121. Quartino A, Huledal G, Sparve E, Luttgen M, Bueters T, Karlsson P, Olsson T, Paraskos J, Maltby J, Claeson-Bohnstedt K, Lee CM, Alexander R, Falting J, Paulsson B (2014) Population pharmacokinetic and pharmacodynamic analysis of plasma $A\beta_{40}$ and $A\beta_{42}$ following single oral doses of the BACE1 inhibitor AZD3839 to healthy volunteers. Clin Pharmacol Drug Dev 3(5):396–405. doi:10.1002/cpdd.130
122. Jacobsen H, Ozmen L, Caruso A, Narquizian R, Hilpert H, Jacobsen B, Terwel D, Tanghe A, Bohrmann B (2014) Combined treatment with a BACE inhibitor and anti-Aβ antibody gantenerumab enhances amyloid reduction in APP_{London} mice. J Neurosci 34(35): 11621–11630. doi:10.1523/JNEUROSCI.1405-14.2014
123. Hoffmann-La Roche. A single-center, randomized, double-blind, single and multiple ascending dose, placebo-controlled study to investigate the safety, tolerability, pharmacokinetics (including the effect of food) and pharmacodynamics of RO5598887 following oral administration in healthy subjects. In: ClinicalTrials.gov [Internet]. Bethesda (MD): National Library of Medicine (US). 2000- [2016 May 02]. Available from: https://clinicaltrials.gov/show/NCT01461967
124. Hoffmann-La Roche. A single-center, randomized, investigator/subject-blind, single dose, placebo-controlled, parallel group study to investigate the pharmacodynamic and pharmacokinetic behavior or RO5508887 in plasma and cerebral spinal fluid following oral administration in healthy volunteers. In: ClinicalTrials.gov [Internet]. Bethesda (MD): National Library of Medicine (US). 2000- [2016 May 02]. Available from: https://clinicaltrials.gov/ct2/show/NCT01592331?term=RO5508887&rank=2
125. Hoffmann-La Roche. A single-center, randomized, investigator/subject-blind, multiple ascending-dose, placebo-controlled study to investigate the safety, tolerability, pharmacokinetics and pharmacodynamics of RO5508887 following oral administration in healthy subjects. In: ClinicalTrials.gov [Internet]. Bethesda (MD): National Library of Medicine (US). 2000- [2016 May 02]. Available from: https://clinicaltrials.gov/show/NCT01664143
126. Pfizer. A phase 1, randomized investigator-and-subject-blind, sponsor open, placebo controlled two-part study to characterize the pharmacokinetics and pharmacodynamics of single doses of PF-05297909 in healthy adult subjects. In: ClinicalTrials.gov [Internet]. Bethesda (MD): National Library of Medicine (US). 2000- [2016 May 02]. Available from: https://clinicaltrials.gov/show/NCT01462851
127. Bell J, O'Neill B, Brodney M, Hajos-Korcsok E, Lu Y, Riddell D, Ito K, Ueckert S, Nicholas T (2013) A novel BACE inhibitor (PF-05297909): a two-part adaptive design to evaluate safety, pharmacokinetics and pharmacodynamics for modifying beta-amyloid in a first-in-human study. Alzheimers Dement 9(4):287. doi:10.1016/j.jalz.2013.05.578
128. High Point Pharmaceuticals LLC. A double-blind, randomized, placebo-controlled, phase I, multiple-dose study to evaluate the safety, tolerability, and pharmacokinetics of orally-administered HPP854 in subjects with mild cognitive impairment or a diagnosis of mild Alzheimer's disease. In: ClinicalTrials.gov [Internet]. Bethesda (MD): National Library of Medicine (US). 2000- [2016 May 02]. Available from: https://clinicaltrials.gov/show/NCT01482013
129. Forman M, Palcza J, Tseng J, Leempoels J, Ramael S, Han D, Jhee S, Ereshefsky L, Tanen M, Laterza O, Dockendorf M, Krishna G, Ma L, Wagner J, Troyer M (2012) The novel BACE inhibitor MK-8931 dramatically lowers cerebrospinal fluid Aβ peptides in healthy subjects following single- and multiple-dose administration. Alzheimers Dement 8(4):P704. doi:10.1016/j.jalz.2012.05.1900
130. Tseng J, Dockendorf M, Krishna G, Ma L, Palcza J, Leempoels J, Ramael S, Han D, Jhee S, Ereshefsky L, Wagner J, Troyer M, Forman M (2012) Safety and pharmacokinetics of the

novel BACE inhibitor MK-8931 in healthy subjects following single- and multiple-dose administration. Alzheimers Dement 8(4 Suppl):P184–P185. doi:10.1016/j.jalz.2012.05.500
131. Merck Sharp & Dohme Corp. A study to assess the safety, tolerability, and pharmacodynamics of MK-8931/SCH 900931 in patients With Alzheimer's disease [phase 1b; protocol No. 010–00 (also known as P07820)]. In: ClinicalTrials.gov [Internet]. Bethesda (MD): National Library of Medicine (US). 2000- [2016 May 02]. Available from: https://clinicaltrials.gov/show/NCT01496170
132. Merck Sharp & Dohme Corp. An open-label, two-part, single-dose study to investigate the pharmacokinetics of MK-8931 in subjects With renal insufficiency (protocol No. MK-8931-009 [P08535]). In: ClinicalTrials.gov [Internet]. Bethesda (MD): National Library of Medicine (US). 2000- [2016 May 02]. Available from: https://clinicaltrials.gov/show/NCT01537757
133. Merck Sharp & Dohme Corp. A randomized, placebo controlled, parallel-group, double blind efficacy and safety trial of MK-8931 with a long term double-blind extension in subjects with mild to moderate Alzheimer's disease (protocol No. MK-8931-017-10) (also known as SCH 900931, P07738). In: ClinicalTrials.gov [Internet]. Bethesda (MD): National Library of Medicine (US). 2000- [2016 May 02]. Available from: https://clinicaltrials.gov/show/NCT01739348
134. Merck (2013) Merck advances development of program for investigational Alzheimer's disease therapy, MK-8931. http://www.mercknewsroom.com/news-release/prescription-medicine-news/merck-advances-development-program-investigational-alzheimer
135. Merck Sharp & Dohme Corp. A phase III, randomized, placebo-controlled, parallel-group, double-blind clinical trial to study the efficacy and safety of MK-8931 (SCH 900931) in subjects with amnestic mild cognitive impairment due to Alzheimer's disease (prodromal AD). In: ClinicalTrials.gov [Internet]. Bethesda (MD): National Library of Medicine (US). 2000- [2016 May 02]. Available from: https://clinicaltrials.gov/study/NCT01953601
136. AstraZeneca. A phase I, randomized, double-blind, placebo-controlled, single ascending dose study to assess the safety, tolerability, pharmacokinetics and effect on biomarkers of AZD3293 including an open-label food effect group in healthy male and non-fertile female volunteers. In: ClinicalTrials.gov [Internet]. Bethesda (MD): National Library of Medicine (US). 2000- [2016 May 02]. Available from: https://clinicaltrials.gov/show/NCT01739647
137. AstraZeneca. A phase I, randomized, double-blind, placebo-controlled, two-part, multiple ascending dose study to assess the safety, tolerability, pharmacokinetics and effect on biomarkers of AZD3293 in plasma and cerebrospinal fluid in healthy male and non-fertile female elderly volunteers and in mild-to-moderate Alzheimer disease patients. In: ClinicalTrials.gov [Internet]. Bethesda (MD): National Library of Medicine (US). 2000- [2016 May 02]. Available from: https://clinicaltrials.gov/show/NCT01795339
138. AstraZeneca. A phase I, single-center, open-label, 3-group, fixed-sequence study to assess the effect of itraconazole, a potent CYP3A4 inhibitor, or diltiazem, a moderate CYP3A4 inhibitor, on the pharmacokinetics of AZD3293 and the effects of AZD3293 on the pharmacokinetics of midazolam, a CYP3A4/CYP3A5 substrate, in healthy young male and female volunteers. In: ClinicalTrials.gov [Internet]. Bethesda (MD): National Library of Medicine (US). 2000- [2016 May 02]. Available from: https://clinicaltrials.gov/show/NCT02010970
139. AstraZeneca. A phase I, randomized, double-blind, placebo-controlled, single and multiple ascending dose study to assess the safety, tolerability, pharmacokinetics and effect on biomarkers of AZD3293 in healthy Japanese male and non-fertile female volunteers including elderly. In: ClinicalTrials.gov [Internet]. Bethesda (MD): National Library of Medicine (US). 2000- [2016 May 02]. Available from: https://clinicaltrials.gov/show/NCT02005211
140. AstraZeneca, Parexel. A single-center, randomized, double-blinded, placebo-controlled, 4-way cross-over study to assess the effect of a single oral dose of AZD3293 administration on QTc interval compared to placebo, using open-label AVELOX (moxifloxacin) as a positive control, in healthy male subjects. In: ClinicalTrials.gov [Internet]. Bethesda (MD):

National Library of Medicine (US). 2000- [2016 May 02]. Available from: https://clinicaltrials.gov/show/NCT02040987
141. AstraZeneca. A phase 1, open-label, randomized, single-dose, 3-period cross-over, relative bioavailability study to assess two solid formulations compared to an oral solution of AZD3293 in healthy male and non-fertile female subjects. In: ClinicalTrials.gov [Internet]. Bethesda (MD): National Library of Medicine (US). 2000- [2016 May 02]. Available from: https://clinicaltrials.gov/show/NCT02039180
142. Eli Lilly and Company, AstraZeneca. A bioequivalence and food effect study in healthy subjects administered 2 different tablet formulations of AZD3293. In: ClinicalTrials.gov [Internet]. Bethesda (MD): National Library of Medicine (US). 2000- [2016 May 02]. Available from: https://clinicaltrials.gov/show/NCT02663128
143. AstraZeneca. A phase I, open-label, single-center study to assess the absorption, metabolism, and excretion after oral administration of [14C]-AZD3293 to healthy male subjects. In: ClinicalTrials.gov [Internet]. Bethesda (MD): National Library of Medicine (US). 2000- [2016 May 02]. Available from: https://clinicaltrials.gov/show/NCT02126514
144. Eli Lilly and Company, AstraZeneca. A study to characterize AZD3293 pharmacokinetics as a function of dosing duration and to determine the effect of AZD3293 on the pharmacokinetics of CYP3A substrates in healthy subjects. In: ClinicalTrials.gov [Internet]. Bethesda (MD): National Library of Medicine (US). 2000- [2016 May 02]. Available from: https://clinicaltrials.gov/show/NCT02406261
145. Eli Lilly and Company, AstraZeneca. Effect of AZD3293 on the pharmacokinetics of warfarin in healthy subjects. In: ClinicalTrials.gov [Internet]. Bethesda (MD): National Library of Medicine (US). 2000- [2016 May 02]. Available from: https://clinicaltrials.gov/show/NCT02540668
146. Eli Lilly and Company, AstraZeneca. Effect of AZD3293 on the pharmacokinetics of dabigatran in healthy subjects. In: ClinicalTrials.gov [Internet]. Bethesda (MD): National Library of Medicine (US). 2000- [2016 May 02]. Available from: https://clinicaltrials.gov/show/NCT02568397
147. AstraZeneca (2014) AstraZeneca and Lilly announce alliance to develop and commercialize BACE inhibitor AZD3293 for Alzheimer's disease. https://www.astrazeneca.com/media-centre/press-releases/2014/astrazeneca-lilly-bace-inhibitor-azd3293-alzheimers-disease-16092014.html
148. Eli Lilly and Company, AstraZeneca. A 24-month, multicenter, randomized, double-blind, placebo-controlled, parallel-group, efficacy, safety, tolerability, biomarker, and pharmacokinetic study of AZD3293 in early Alzheimer's disease (The AMARANTH Study). In: ClinicalTrials.gov [Internet]. Bethesda (MD): National Library of Medicine (US). 2000- [2016 May 02]. Available from: https://clinicaltrials.gov/show/NCT02245737
149. Janssen Research & Development LLC. A double-blind, placebo-controlled, randomized, single-ascending dose study to investigate the safety, tolerability and pharmacokinetics of JNJ-54861911 in healthy subjects. In: ClinicalTrials.gov [Internet]. Bethesda (MD): National Library of Medicine (US). 2000- [2016 May 02]. Available from: https://clinicaltrials.gov/show/NCT01827982
150. Janssen Research & Development LLC. A double-blind, placebo-controlled, randomized, multiple-ascending dose study to investigate the safety, tolerability, pharmacokinetics and pharmacodynamics of JNJ-54861911 in healthy elderly subjects. In: ClinicalTrials.gov [Internet]. Bethesda (MD): National Library of Medicine (US). 2000- [2016 May 02]. Available from: https://clinicaltrials.gov/show/NCT01887535
151. Janssen Research & Development LLC. A randomized, open-label, 3-Way crossover study in healthy older male subjects to evaluate the bioavailability, food effect, safety and tolerability of a solid dosage formulation of JNJ-54861911. In: ClinicalTrials.gov [Internet]. Bethesda (MD): National Library of Medicine (US). 2000- [2016 May 02]. Available from: https://clinicaltrials.gov/show/NCT02260700

152. Janssen Research & Development LLC. A double-blind, placebo-controlled, randomized, 4-week, multiple-dose, proof-of-mechanism study in subjects with prodromal Alzheimer's disease investigating the effects of JNJ-54861911 on Aβ processing in CSF and plasma. In: ClinicalTrials.gov [Internet]. Bethesda (MD): National Library of Medicine (US). 2000- [2016 May 02]. Available from: https://clinicaltrials.gov/show/NCT01978548
153. Janssen Pharmaceutical K.K. A double-blind, placebo-controlled, randomized, single-ascending dose study to investigate the safety, tolerability and pharmacokinetics of JNJ-54861911 in healthy Japanese male subjects. In: ClinicalTrials.gov [Internet]. Bethesda (MD): National Library of Medicine (US). 2000- [2016 May 02]. Available from: https://clinicaltrials.gov/show/NCT02180269
154. Janssen Pharmaceutical K.K. A double-blind, placebo-controlled, randomized, 4-Week, multiple-dose, proof of mechanism (POM) study in Japanese subjects asymptomatic at risk for Alzheimer dementia (ARAD) investigating the effects of JNJ-54861911 on Aβ Processing in cerebrospinal fluid (CSF) and plasma. In: ClinicalTrials.gov [Internet]. Bethesda (MD): National Library of Medicine (US). 2000- [2016 May 02]. Available from: https://clinicaltrials.gov/show/NCT02360657
155. Janssen Research & Development LLC. An open-label, fixed-sequence study to assess effects of clarithromycin on the single-dose pharmacokinetics of JNJ-54861911 in healthy male subjects. In: ClinicalTrials.gov [Internet]. Bethesda (MD): National Library of Medicine (US). 2000- [2016 May 02]. Available from: https://clinicaltrials.gov/show/NCT02197884
156. Janssen Research & Development LLC. An open-label, fixed-sequence study in healthy male subjects to assess the drug interaction potential of multiple-doses of JNJ-54861911 with a drug "cocktail" representative for CYP3A4, CYP2B6, CYP2C9, and CYP1A2 substrates. In: ClinicalTrials.gov [Internet]. Bethesda (MD): National Library of Medicine (US). 2000- [2016 May 02]. Available from: https://clinicaltrials.gov/show/NCT02211079
157. Janssen Research & Development LLC. A phase 1, 2-panel, open-label, fixed-sequence study in healthy adult subjects to investigate the pharmacokinetic interaction between JNJ-54861911 and transporter substrates rosuvastatin and metformin. In: ClinicalTrials.gov [Internet]. Bethesda (MD): National Library of Medicine (US). 2000- [2016 May 02]. Available from: https://clinicaltrials.gov/show/NCT02611518
158. Janssen Research & Development LLC. A randomized, double-blind, placebo- and positive-controlled, multiple-dose, four-way, cross-over study to evaluate the effects of repeated oral doses of JNJ-54861911 on electrocardiogram intervals in healthy subjects. In: ClinicalTrials.gov [Internet]. Bethesda (MD): National Library of Medicine (US). 2000- [2016 May 02]. Available from: https://clinicaltrials.gov/show/NCT02152332
159. Janssen Research & Development LLC. An open-label, randomized, three-period crossover study to evaluate the relative oral bioavailability and food effect of JNJ-54861911 tablet (1x25mg) after single dose administration in healthy elderly subjects. In: ClinicalTrials.gov [Internet]. Bethesda (MD): National Library of Medicine (US). 2000- [2016 May 02]. Available from: https://clinicaltrials.gov/show/NCT02355561
160. Janssen Research & Development LLC. A phase 2a randomized, double-blind, placebo-controlled, parallel-group, multi-center study investigating the safety and tolerability of JNJ-54861911 in subjects with early Alzheimer's disease. In: ClinicalTrials.gov [Internet]. Bethesda (MD): National Library of Medicine (US). 2000- [2016 May 02]. Available from: https://clinicaltrials.gov/show/NCT02260674
161. Janssen Research & Development LLC. A randomized, two-period, double-blind placebo-controlled and open-label, multicenter extension study to determine the long-term safety and tolerability of JNJ-54861911 in subjects in the early Alzheimer's disease spectrum. In: ClinicalTrials.gov [Internet]. Bethesda (MD): National Library of Medicine (US). 2000- [2016 May 02]. Available from: https://clinicaltrials.gov/show/NCT02406027
162. Janssen Research & Development LLC. A phase 2b/3 randomized, double-blind, placebo-controlled, parallel group, multicenter study investigating the efficacy and safety of JNJ-54861911 in subjects who are asymptomatic at risk for developing Alzheimer's

dementia. In: ClinicalTrials.gov [Internet]. Bethesda (MD): National Library of Medicine (US). 2000- [2016 May 02]. Available from: https://clinicaltrials.gov/show/NCT02569398
163. Boehringer Ingelheim. Safety, tolerability, pharmacokinetics and pharmacodynamics of single rising oral doses of BI 1181181 in healthy male volunteers in a partially randomised, single-blind, placebo-controlled trial, and investigation of relative bioavailability and the effect of food on the pharmacokinetics of BI 1181181(open-label, randomised, three-way cross-over design). In: ClinicalTrials.gov [Internet]. Bethesda (MD): National Library of Medicine (US). 2000- [2016 May 02]. Available from: https://clinicaltrials.gov/show/NCT02044406
164. Boehringer Ingelheim. Safety, tolerability, pharmacokinetics and pharmacodynamics of single oral doses of BI 1181181 in young healthy male volunteers (randomised, double-blind, placebo-controlled within dose groups phase I trial). In: ClinicalTrials.gov [Internet]. Bethesda (MD): National Library of Medicine (US). 2000- [2016 May 02]. Available from: https://clinicaltrials.gov/show/NCT02106247
165. Vitae Pharmaceuticals (2014) Positive top-line results achieved from two phase 1 clinical trials of BACE inhibitor BI1181181/VTP-37948 in Alzheimer's disease. http://ir.vitaepharma.com/phoenix.zhtml?c=219654&p=irol-newsArticle&ID=1981037
166. Boehringer Ingelheim. Safety, tolerability, pharmacokinetics, and pharmacodynamics of multiple rising doses of BI 1181181 given orally q.d. for 10 days in young healthy male and elderly healthy male/female volunteers (randomized, double-blind, placebo controlled within dose groups phase I study). In: ClinicalTrials.gov [Internet]. Bethesda (MD): National Library of Medicine (US). 2000- [2016 May 02]. Available from: https://clinicaltrials.gov/show/NCT02254161
167. Vitae Pharmaceuticals (2015) BACE inhibitor BI 1181181 voluntarily put on temporary clinical hold for safety evaluation. http://ir.vitaepharma.com/phoenix.zhtml?c=219654&p=irol-newsArticle&ID=2020814
168. Boehringer Ingelheim. A study to investigate the effects of BI 1181181 on the pharmacokinetics of midazolam, warfarin, omeprazole and digoxin in healthy male subjects. In: ClinicalTrials.gov [Internet]. Bethesda (MD): National Library of Medicine (US). 2000- [2016 May 02]. Available from: https://clinicaltrials.gov/show/NCT02345304
169. Novartis Pharmaceuticals. A randomized, double-blind, placebo-controlled, parallel-group study to assess the safety, tolerability, pharmacokinetics and pharmacodynamics of multiple oral doses of CNP520 in healthy elderly subjects. In: ClinicalTrials.gov [Internet]. Bethesda (MD): National Library of Medicine (US). 2000- [2016 May 03]. Available from: https://clinicaltrials.gov/show/NCT02576639
170. Novartis Pharmaceuticals, Banner Alzheimer's Institute, National Institute on Aging, Alzheimer's Association, Amgen. A randomized, double-blind, placebo-controlled, two-cohort, parallel group study to evaluate the efficacy of CAD106 and CNP520 in participants at risk for the onset of clinical symptoms of Alzheimer's disease. In: ClinicalTrials.gov [Internet]. Bethesda (MD): National Library of Medicine (US). 2000- [2016 May 03]. Available from: https://clinicaltrials.gov/show/NCT02565511

Top Med Chem (2017) 24: 87–118
DOI: 10.1007/7355_2016_19

Published online: 15 March 2017

γ-Secretase Modulators as Aβ42-Lowering Pharmacological Agents to Treat Alzheimer's Disease

Douglas S. Johnson and Martin Pettersson

Abstract γ-Secretase is an intramembrane aspartyl protease comprised of four essential subunits including presenilin, nicastrin (NCT), anterior pharynx defective 1 (Aph-1), and presenilin enhancer 2 (Pen-2). The amyloid precursor protein (APP) is cleaved sequentially by β-secretase and γ-secretase to generate Aβ peptides including neurotoxic Aβ42 monomers and oligomers that are believed to be key pathological species in AD. Familial Alzheimer's disease (FAD) mutations in presenilin and APP increase the relative proportion of Aβ42. γ-Secretase modulators (GSMs) have been discovered that bind to presenilin and selectively modulate γ-secretase proteolytic activity. Importantly, GSMs have the opposite effect on the Aβ cleavage profile as compared to FAD mutations, namely they decrease the relative proportion of Aβ42. This review will discuss the initial discovery of GSMs and the recent progress leading to the development of GSMs with improved drug-likeness. These efforts have culminated in GSMs that are currently undergoing proof-of-mechanism studies in the clinic, which is a significant step forward toward testing the amyloid hypothesis.

Keywords Aβ42, Alzheimer's disease, Amyloid precursor protein, Medicinal chemistry, Presenilin, γ-Secretase, γ-Secretase modulator

Contents

D.S. Johnson (✉) and M. Pettersson
Pfizer Worldwide Research and Development, 610 Main Street, Cambridge, MA 02139, USA
e-mail: doug.johnson@pfizer.com

1 Introduction

Alzheimer's disease (AD) is a progressive, neurodegenerative disorder and is the most common form of dementia. It results in difficulty remembering newly learned information, memory loss, disorientation, mood and behaviour changes, and eventually difficulty speaking and walking. The pathological hallmarks of AD include deposition of amyloid-β peptides (Aβ) in the brain parenchyma [1], formation of paired helical filaments (PHFs) of tau aggregates [2], and ultimately neuronal loss. Genetic linkage studies and positional cloning efforts revealed that rare mutations in amyloid precursor protein (APP), PSEN1, and PSEN2 cause early onset familial AD (EO-FAD) [3]. The vast majority of these mutations (>200) increase the Aβ42/Aβ40 ratio, and this relative increase in Aβ42 may promote aggregation leading to oligomers and amyloid fibrils. This convergence of disease pathology and genetics led to the amyloid cascade hypothesis, which suggests that a central event leading to AD is the aggregation of amyloid-β peptides (Aβ), especially Aβ42, which forms toxic soluble oligomers and deposited Aβ in amyloid plaques [4, 5]. Therefore pharmacological approaches to reduce the levels of Aβ peptides have garnered much attention in the field [6–8].

APP is initially cleaved by β-secretase BACE1 to generate sAPPβ and β-CTF (also referred to as C99), which is cleaved by γ-secretase in the membrane to form APP intracellular domain (AICD) and multiple Aβ peptides (Fig. 1). γ-Secretase initially cleaves APP β-CTF at the ε-site at the membrane/cytosolic border to generate Aβ49 and/or Aβ48, which undergo successive stepwise cleavage every three to four amino acids to generate shorter forms which are released from the

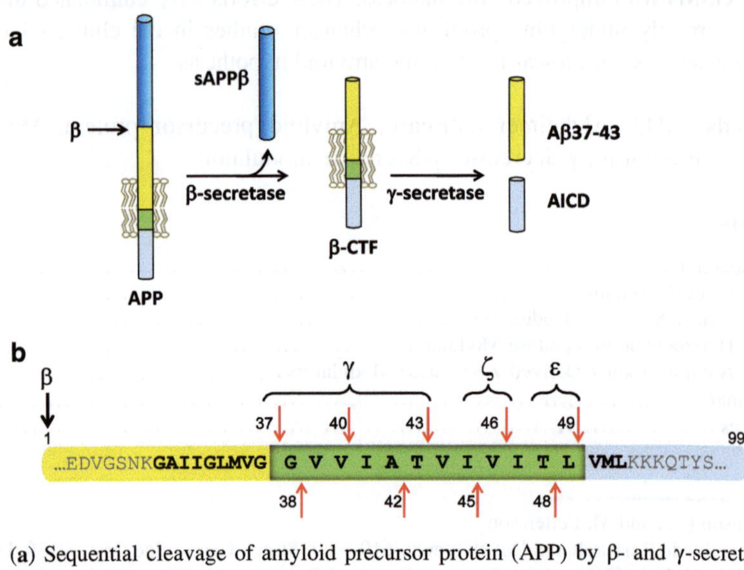

Fig. 1 (**a**) Sequential cleavage of amyloid precursor protein (APP) by β- and γ-secretases and corresponding products. (**b**) Sequence of the membrane and nearby regions of the APP β-CTF substrate. *Vertical red arrows* show locations of γ, ζ, and ε cleavages by γ-secretase

membrane and secreted. It has been proposed that γ-secretase cleavage occurs along two product lines with Aβ49 leading to Aβ46, Aβ43, Aβ40, and Aβ37, and Aβ48 leading to Aβ45, Aβ42, and Aβ38 [9]; however, this is probably oversimplified as recent studies have shown that the product lines can cross and the initial γ-secretase cleavage does not always define subsequent product lines [10–12].

There are several lines of evidence that suggest that Aβ42 is a key toxic species contributing to AD. First, although Aβ42 is a minor Aβ species, it is the earliest and most predominant form deposited in amyloid plaques in humans [13] and transgenic mouse models [14, 15]. In part, this is because Aβ42 is more hydrophobic and more prone to aggregation [16]. Consistent with this early work, recent atomic force microscopy studies show that oligomers of Aβ42, but not of Aβ40, seed fibril formation [17]. In addition to the importance of Aβ42, it is also important to consider the ratio of longer Aβ species to shorter Aβ species, where the Aβ42/Aβ40 ratio has received the most attention to date (although with the advent of γ-secretase modulators (GSMs) that elevate Aβ38 and/or Aβ37, these species will need to be factored in as well). Analysis of the Aβ profiles for several familial Alzheimer's disease (FAD) PS mutations suggests that the Aβ42/Aβ40 ratio, rather than the total amount of Aβ, correlates better with the age of onset of FAD [18]. It has been suggested that a potential reason for the importance of the Aβ42/Aβ40 ratio could be that Aβ40 plays a protective role by sequestering the more toxic Aβ42, thereby preventing further oligomerization of Aβ42. In fact, several in vitro studies support this hypothesis and show that Aβ40 can inhibit the aggregation of Aβ42 in an Aβ42/Aβ40 ratio-dependent manner [19, 20]. Other studies indicate that Aβ40 and Aβ42 influence their respective aggregation properties where Aβ40 inhibits Aβ42 aggregation and Aβ42 stimulates Aβ40 aggregation [21–23]. Also small increases in the Aβ42/Aβ40 ratio have been shown to exhibit increased synaptotoxicity. In addition, transgenic models that express Aβ40 or Aβ42 have been developed, and mice expressing Aβ42 accumulate insoluble Aβ42 and develop amyloid plaques, whereas the mice expressing Aβ40 do not develop amyloid pathology [24]. Furthermore, Aβ42, but not Aβ40, has been shown to induce tau phosphorylation, aggregation, and cleavage [25].

The γ-secretase complex is an intramembrane aspartyl protease and consists of four essential subunits including presenilin (PS), nicastrin (NCT), presenilin enhancer 2 (Pen-2) and anterior pharynx defective 1 (Aph-1) [26]. PS is the catalytic subunit and during assembly of the complex, it undergoes endoproteolysis to produce a heterodimer composed of PS1-NTF and PS1-CTF, which constitutes the active form of the enzyme. An atomic structure of human γ-secretase has been determined by cryo-electron microscopy (Fig. 2) [27, 28]. This structure allows visualization of all 20 transmembrane domains and the interactions between the four components that make up the γ-secretase complex. γ-Secretase has been a highly sought-after target to inhibit Aβ production, and there has been considerable success developing potent brain-penetrant γ-secretase inhibitors (GSI) with several advancing to human clinical trials. Unfortunately, both semagacestat [29] and avagacestat [30] failed in Phase III and Phase II clinical trials, respectively, and patients receiving the GSIs actually trended toward a worsening of cognition and increased incidence of gastrointestinal adverse events and skin cancer compared with those who received placebo [31]. γ-Secretase has many

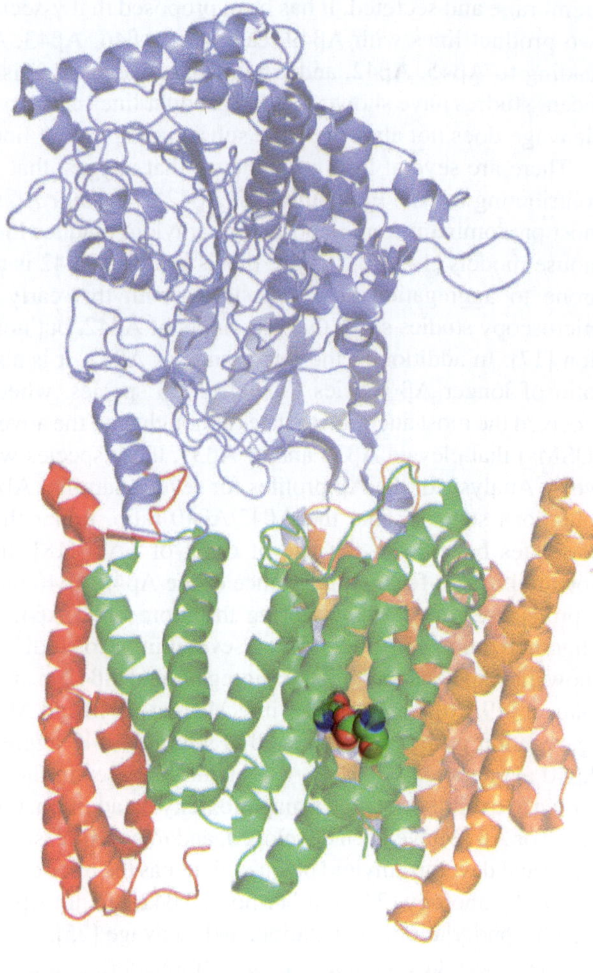

Fig. 2 The γ-secretase complex cryo-EM structure (5FN2) [27]. Nicastrin (NCT) is shown in *blue*, anterior pharynx defective 1 (Aph-1) in *orange*, PS1 in *green* and presenilin enhancer 2 (Pen-2) in *red*. The active site aspartic acids are shown as *spheres*

substrates in addition to APP [32], most notably Notch, and it is thought that at least some of the side effects result from inhibition of Notch cleavage [33]. In addition, γ-secretase inhibition results in accumulation of the β-CTF of APP, and this has been proposed as another potential reason for the detrimental side effects of GSIs [34]. Given the safety concerns with inhibition of γ-secretase, alternative approaches to modulate γ-secretase activity have been a topic of significant interest and will be described in the next section.

2 γ-Secretase Modulators

GSMs were discovered in 2001 when a subset of non-steroidal anti-inflammatory drugs (NSAIDs) were found to specifically lower Aβ42 and increase Aβ38, albeit with weak potency [35]. The first generation of NSAID GSMs include sulindac sulfide, indomethacin, (*R*)-ibuprofen, and (*R*)-flurbiprofen (tarenflurbil) (Fig. 3). The Aβ42 lowering was independent of the cyclooxygenase activity [35, 36] and *R*-flurbiprofen selectively lowered Aβ42 levels in a broken cell γ-secretase assay (Aβ42 IC_{50} ~200–300 μM) suggesting a direct but weak interaction with γ-secretase [37]. Importantly, these NSAIDs did not inhibit the cleavage of substrates, and the intracellular domains of APP (AICD) and Notch (NICD) were still formed. Despite the weak potency and poor brain penetration, tarenflurbil was advanced into clinical studies, but it did not slow cognitive decline in patients with mild AD in a phase 3 clinical trial [38, 39].

GSMs have several unique and attractive features that suggest that they could be potential therapeutics for AD [7]. They cause: (1) selective decrease of Aβ42 or simultaneous decrease of Aβ42 and Aβ40; (2) increase in shorter Aβ species (Aβ38/Aβ37); although the implication of elevating Aβ37 or Aβ38 in AD is unknown, it is believed that the shorter forms are less prone to aggregation and less pathogenic than Aβ42 [17]; (3) no significant effect on the total amount of Aβ produced; (4) no accumulation of APP β-CTF; (5) no plasma Aβ rebound; and (6) preserve ε cleavage of γ-secretase substrates and, therefore, do not affect the production of APP- and Notch-ICDs (AICD and NICD). Notably, the effect of GSMs on Aβ peptides is opposite that of many FAD presenilin mutations [40–43], which increase the relative proportion of Aβ42 [3]. Furthermore, GSMs increase the levels of shorter peptides (i.e. Aβ37 and Aβ38), and there is some evidence suggesting that these shorter species could actually inhibit the aggregation of Aβ42. Therefore, many pharmaceutical companies have invested heavily in this area, and several different classes of GSMs have been discovered as potential disease-modifying agents for the treatment of AD. These classes can be divided into three categories: NSAID-derived acid GSMs, non-NSAID heterocyclic GSMs, and natural product-derived GSMs as described below. There is some controversy whether NSAID GSMs bind to γ-secretase and/or to the substrate, APP [44, 45], but chemical biology studies

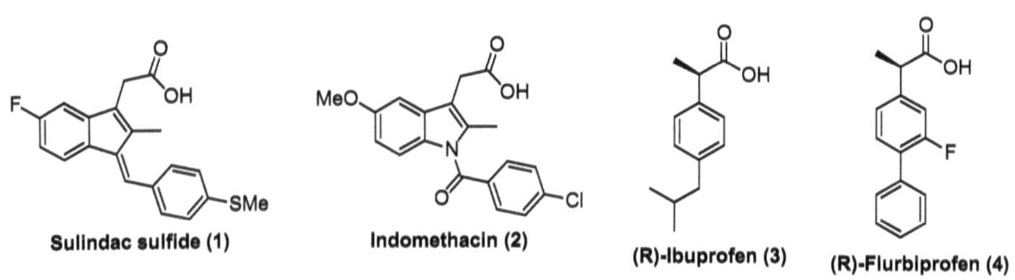

Fig. 3 Non-steroidal anti-inflammatory drug (NSAID) γ-secretase modulators (GSMs)

have shown that second generation GSMs bind directly to PS1-NTF [46–52], although the different classes appear to have distinct binding sites [51].

2.1 Acid γ-Secretase Modulators

Chiesi has developed a series of flurbiprofen analogs with improved potency for lowering Aβ42. This culminated in CHF5074 (**5**) which lowered Aβ42 with an IC_{50} between 3 and 41 μM (Fig. 4) [53]. In a Phase I study looking at the pharmacokinetics and pharmacodynamics of CHF5074 (200, 400, and 600 mg/day), no lowering of Aβ levels in both plasma and cerebrospinal fluid (csf) was detected. However, it was found that the soluble CD40 ligand, a marker of microglia activation, was lowered in a dose-dependent fashion [54]. So it appears that the compound may be acting as a microglial modulator rather than a GSM. Because of the weak potency and poor brain penetration of early NSAID-derived compounds, a key goal has been to develop GSM compounds with improved potency and brain availability. Several additional groups have also explored this chemical space and were able to generate compounds with Aβ42 IC_{50s} less than 200 nM by adding additional substituents to the phenyl acetic acid core of flurbiprofen (Fig. 4). For example, EnVivo has reported EVP-0015962 (**6**) which decreased Aβ42 with an IC_{50} of 67 nM [55]. Consistent with the profile of a GSM, Aβ38 was increased while Aβ total, APP β-CTF, and AICD/NICD generation were not changed. Chronic treatment of EVP-0015962 in Tg2576 mice (20 and 60 mg/kg/day) resulted in brain concentrations of 2.5 and 8.3 μM, respectively, with corresponding

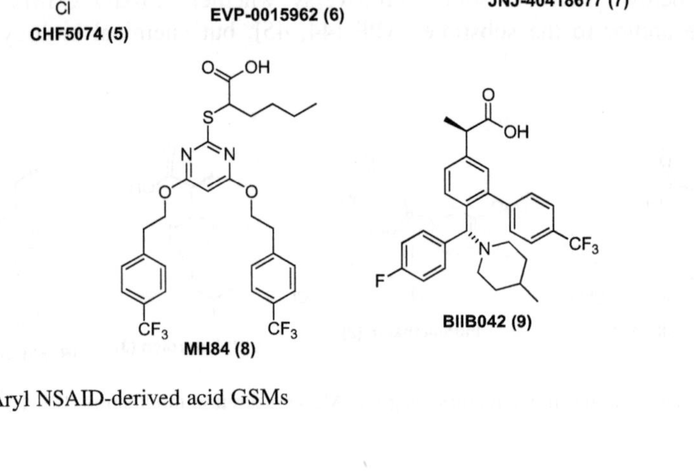

Fig. 4 Aryl NSAID-derived acid GSMs

reduction of cytosolic, membrane-bound, and aggregated Aβ42. In addition, amyloid deposition and cognitive deficits were reduced. Janssen added a second trifluoromethylphenyl substituent to give JNJ-40418677 (**7**), a brain-penetrant phenyl acetic acid GSM. JNJ-40418677 reduced Aβ42 secretion in cells and primary rat cortical neuronal cultures with an IC_{50} of approximately 200 nM and did not change levels of total Aβ or Notch processing [56]. Acute treatment with JNJ-40418677 selectively reduced brain Aβ42 levels in wild-type mice with a concomitant increase in Aβ38. Chronic administration in Tg2576 mice from 6 to 13 months of age resulted in dose-dependent reduction in amyloid plaque load. Intriguingly, all Aβ species in soluble and deposited fractions were reduced, which deviates from a typical GSM profile and merits further investigation. Another series of aryl carboxylic acid GSMs, exemplified by MH84 (**8**), has been shown to have PPARγ agonist activity [57]. However, these compounds are very lipophilic and the potency is weak so they have not advanced beyond being tool compounds. Biogen has also published on a series of aryl acid GSMs that merge the core of flurbiprofen with features present in GSM-1 (**10**, Fig. 5). For example, BIIB042 (**9**) lowered Aβ42 in CHO cells with an IC_{50} of 170 nM and increased Aβ38 with an EC_{50} of 150 nM [58]. A reduction of brain Aβ42 was observed when BIIB042 was administered to mice, rats, and cynomolgus monkeys.

Merck made a significant advance by replacing the core aryl ring of previous NSAID-derived GSMs with a piperidine ring (Fig. 5). Then they optimized the substituents to generate a potent series of piperidine acetic acid GSMs, exemplified by GSM-1 (**10**) [59, 60]. GSM-1 has an Aβ42 IC_{50} value of 120–348 nM and selectively reduced Aβ42 and increased Aβ38, while the levels of total Aβ and AICD did not

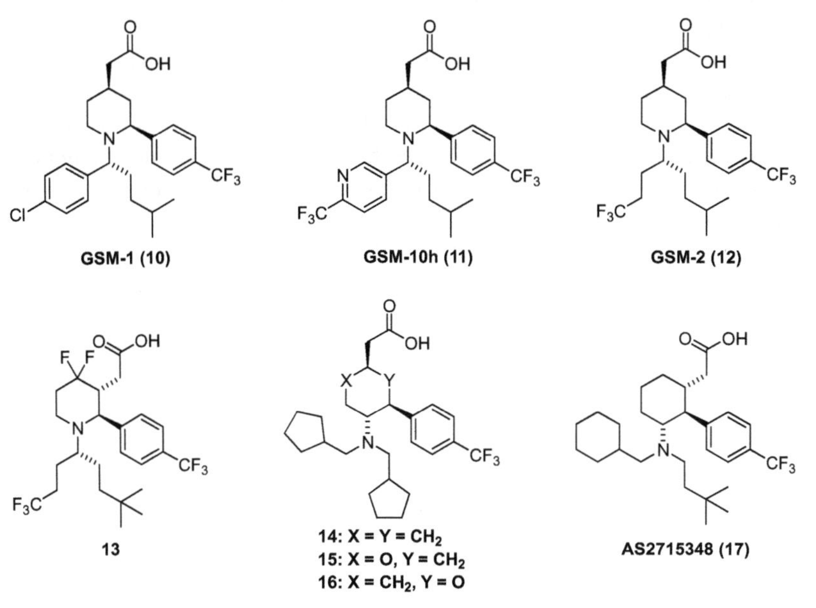

Fig. 5 Piperidine and aminocyclohexane acid GSMs

change. More recently, GSK reported GSM-10h (**11**), which is a pyridyl analog of GSM-1 with lower lipophilicity and similar potency (Aβ42 IC_{50} = 300 nM) [61]. GSM-10h increased Aβ38 and had no effect on total Aβ levels indicating that the rise in Aβ38 compensates for the decrease in Aβ42. GSM-10h demonstrated excellent bioavailability and good central nervous system (CNS) penetration and decreased Aβ42 in plasma, CSF, and brain of rats. In contrast to GSIs, GSM-10h did not result in accumulation of β-C-terminal fragment (β-CTF) and did not cause Aβ rebound [62]. In addition, Astellas has reported the pharmacological profile of another piperidine acetic acid GSM referred to as GSM-2 (**12**) [34]. GSM-2 selectively lowered Aβ42 with an IC_{50} of 65 nM, while increasing Aβ38 (EC_{50} = 81 nM) with no effect on β-CTF levels or Notch signalling. In contrast to the GSIs LY450139 and BMS-708,163, administration of GSM-2 for 8 days ameliorated the cognitive deficit in 5.5-month-old Tg2576 mice as measured in the Y-maze test. All three drugs reduced Aβ42 levels, but only the two GSIs increased β-CTF levels and some have suggested that β-CTF elevation could partially explain the cognitive impairment associated with GSI treatment. In addition, 8-day administration of GSM-2 to 18-month-old Tg2576 mice reduced newly synthesized soluble Aβ42 (measured using stable isotope-labeling) and significantly ameliorated cognitive deficits in the Y-maze test [63]. Brain soluble Aβ42 was not reduced, suggesting that newly synthesized Aβ42 contributes more to the cognitive deficits in Tg2576 mice than plaque-associated soluble Aβ42. Merck has also disclosed a series of fluorinated piperidine acetic acid GSMs, including the isomeric 1,2,3-trisubstituted piperidine **13** [64].

Pfizer and Astellas have disclosed a variation of the piperidine acetic acid GSM series that involves transposing the piperidine nitrogen out of the ring providing an aminocyclohexane. This was done in part to alleviate the concern of metabolic activation of the piperidine ring system and subsequent hepatotoxicity that has been observed for some piperidine carboxylic acid derivatives [65]. The amino group was substituted with various alkyl and cycloalkyl groups to give potent GSMs [66]. For example, the bis (cyclopentylmethyl)amino analog **14** has an Aβ42 IC_{50} of 229 nM. It was also possible to incorporate an oxygen in the ring to give the corresponding aminotetrahydropyrans. The 2,5,6-trisubstituted tetrahydropyran **15** and 2,4,5-trisubstituted tetrahydropyran **16** were reported to reduce Aβ42 levels in cells with an IC_{50} of 357 and 777 nM, respectively. This modification is potentially attractive because it lowers the lipophilicity of the compounds and decreases the basicity of the nitrogen. Furthermore, Astellas extended the series to include 1,2,3-trisubstituted aminocyclohexanes with the analogous substitution pattern as piperidine **13** [67]. An example is AS2715348 (**17**), which has been reported to lower Aβ42 levels in various cell types with an IC_{50} between 62 and 156 nM and did not affect β-CTF levels or Notch signalling. In addition, AS2715348 selectively reduced Aβ42 in the brain of rats, Tg2576 mice, and non-human primates. It remains to be seen if an acceptable safety profile with adequate brain exposure at the target can be achieved with the class of acid GSMs.

2.2 *Heterocyclic γ-Secretase Modulators*

The second major class of GSMs is characterized by an A–B–C–D-type heterocyclic scaffold in which the A–B subunit often features an arylimidazole moiety. This series was first disclosed by NeuroGenetics and was further refined by Eisai [68, 69]. NeuroGenetics recently reported on the discovery of **18** [70], which has been identified as NGP-555 (Fig. 6). This series originated from a high-throughput screen of an 80,000 compound library that was assembled from commercial sources. An initial screening hit with an Aβ42 IC_{50} of 15 μM was optimized through a series of targeted libraries and follow-up analogues to arrive at **18**. This lead compound was reported to have an Aβ42 IC_{50} of 10 nM in SH-SY5Y APP cells, and 29 nM in Tg2576 MBCs. Unlike acid GSMs, which selectively reduce Aβ42, the A–B–C–D-type heteroaryl series reduces both Aβ42 and Aβ40 while increasing the levels of Aβ38 and Aβ37, as confirmed by immunoprecipitation mass spectrometry experiments. Four-month-old Tg2576 mice were dosed with **18** at 5–100 mg/kg p.o. for three consecutive days, and brain and plasma samples were analysed 6 h post-dose. Dose-dependent reductions of Aβ42 and Aβ40 were observed, and a 50 mg/kg dose afforded approximately a 30% reduction of brain Aβ42. Next, **18** was tested in a 7-month plaque reduction study in Tg2576 mice in which the compound was administered via the food pellets to achieve a plasma concentration of 3.95 μM [70]. Serial extraction of brain samples was performed to analyse the levels of soluble, detergent-soluble, and formic acid-soluble Aβ peptides. Not only was a significant reduction of Aβ42 and Aβ40 observed in all three fractions, but a robust reduction in Aβ38 was observed as well. This result stands in sharp contrast to observations in the acute in vivo studies where the levels of Aβ38 increase. The

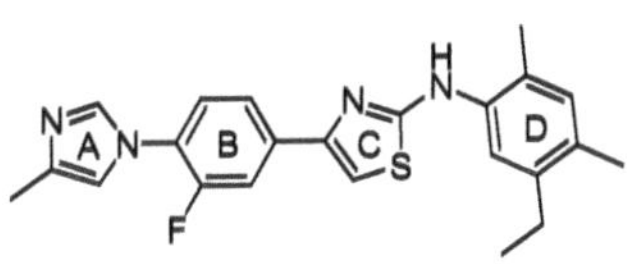

NGP-555 (18)

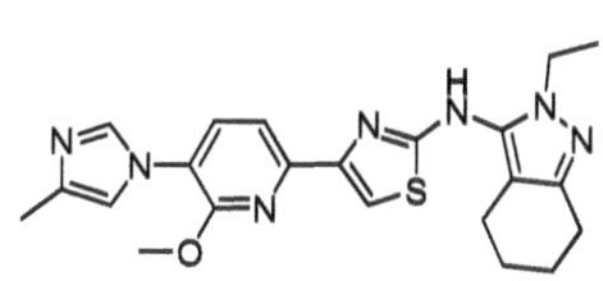

19

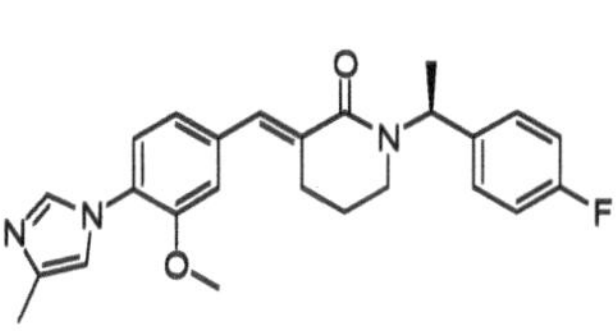

E2012 (20)

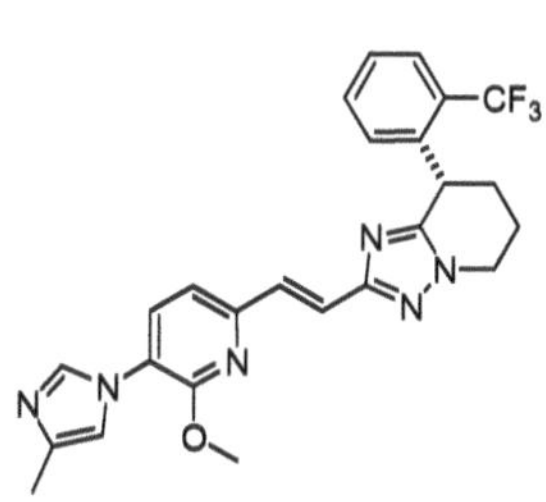

E2212: putative structure (21)

Fig. 6 NeuroGenetics and Eisai GSMs

reason for the reduction of Aβ38 levels in the plaque reduction study is not understood at this point.

The physicochemical properties of **18** and related compounds required further optimization. In particular, the kinetic solubility of **18** was found to be very low (<0.1 μM), which may be attributed to high lipophilicity (*c*logP = 7.7) in combination with four aromatic rings [71]. A recent report described efforts to improve solubility by replacing two of the phenyl rings with heterocycles, and the exemplified compounds have *c*logP values ranging from 3.3 to 5.7 and kinetic solubility <1.6–18.4 μM [72]. Compound **19** incorporating a methoxypyridine as the B-ring and a fused pyrazole in place of the D-ring was the most potent analogue with an Aβ42 IC_{50} of 30 nM (*c*logP = 5.5, kinetic solubility = 2.7 μM). Although **19** is an improvement relative to **18**, it highlights the significant challenge of balancing sufficient potency with adequate physicochemical properties within this chemotype, and this problem is evident throughout the GSM literature (vide infra). Nevertheless, in spite of the challenges encountered in this series, NGP-555 was recently reported to enter Phase I clinical development [73].

In 2005, Eisai published a patent application that claimed compounds such as E2012 (**20**) featuring a lactam as the C-ring hydrogen-bond acceptor and with reduced aromatic ring count relative to **18** and **19** (*c*logP = 4.8) [69]. E2012 has been extensively profiled in vitro and in vivo: Aβ42-lowering activity was determined in several cell lines including HEK $293^{\text{Swe-Lon}}$ cells [74], rat cortical neuronal cultures [75], and CHO-APP cells [76], and the IC_{50} values were 64, 92, and 143 nM, respectively. In vivo reduction of Aβ42 has been assessed in rodent and dog CSF biomarker studies, and robust and dose-dependent reduction of brain Aβ42 has been reported [77, 78]. A Phase I trial with E2012 began in 2006, marking the first non-NSAID GSM to enter clinical development [79]. However, it was subsequently put on hold due to the observation of lenticular opacity (cataract) in a 13-week rat safety study. A recent paper linked the ocular toxicity to increased levels of lenticular desmosterol resulting from inhibition of DHCR24, which is the enzyme responsible for conversion of desmosterol to cholesterol [80]. Elevation of plasma desmosterol was demonstrated to be an effective toxicity biomarker. Furthermore, cataracts were not observed in a 13-week monkey study, and the clinical trial was therefore allowed to proceed. E2012 was tested at doses of 1–400 mg in a single ascending dose study, and plasma Aβ42 was reduced by ~50% at the 400 mg dose [79].

Further development of E2012 was not pursued, and focus was rather shifted to the more potent clinical candidate E2212 (**21**; putative structure). This compound has an IC_{50} of 9 nM in rat cortical neuronal cultures (*c*logP = 4.8), and it was reported to have an improved safety profile [81]. Robust reductions of brain Aβ42 have been demonstrated in multiple preclinical species. In a rat in vivo efficacy study, brain Aβ42 was reduced by 47.7% and 70% following doses of 15 and 100 mg/kg p.o., respectively [78]. This novel GSM retains the olefin linker of E2012, but the central lactam has been replaced with a fused triazole moiety. It is interesting to note that **21** incorporates conformational constraint in the C–D region, and this has become a reoccurring theme in the subsequent patent literature (vide infra). E2212 was evaluated in a single ascending dose study in healthy human volunteers, and it was found to be well tolerated following single doses up to 250 mg. It has a terminal half-life ranging from 12.5 to 19 h, and a 53.6% reduction in plasma Aβ42 was achieved following the 250 mg dose [81]. However, further clinical development of E2212 has not been reported.

The discovery and initial optimization of the A–B–C–D-type heteroaryl series by NeuroGenetics and Eisai generated substantial interest throughout the industry, and these chemotypes have served as starting points for numerous GSM programs (vide infra) [82–85]. Analysis of the literature reveals that the A–B region is highly conserved whereas much greater structural diversity has been achieved within the C- and D-ring sections of this class of GSMs. The frequently used A-ring imidazole appears to be making a key polar interaction with the target binding site. Recently, the imidazole has successfully been replaced with a variety of other heterocycles that can maintain the polar interaction including triazole, pyrazole, oxazole, and pyridyl [83]. The B-ring functions as a rigid spacer to link the A- and the C-rings together within a defined spacial configuration. The C-ring makes a critical polar interaction with the enzyme presumably serving as a hydrogen-bond acceptor; however, there is considerable flexibility as to what types of polar functionality are tolerated in this region. Examples include a wide variety of heterocycles as well as carbonyl-containing scaffolds. Finally, the D-ring appears to be directed into a lipophilic pocket, and the vast majority of GSMs incorporate a lipophilic aromatic ring that is substituted with halogens or other non-metabolizable lipophilic substituents. Only recently have GSMs with robust Aβ42-lowering activity appeared in the literature where this aromatic ring has been replaced with aliphatic systems [83].

The GSM program at Schering-Plough (now Merck) has developed several alternative chemotypes using E2012 as the starting point. Many creative C-ring replacements have been examined including imino heterocycles [86], and 5,6-bicyclic heterocycles [87], the latter conceptually arising from cyclizing the carbonyl of E2012 onto the olefin linker. These efforts ultimately led to the discovery a pyrazolopyridine series as exemplified by **22** (Aβ42 IC_{50} = 107 nM, *c*logP = 5.4, Fig. 7) [88]. When dosed in rat at 30 mg/kg p.o., this compound achieved a 45% reduction of CSF Aβ42 at the 3-h time point, and the corresponding brain and plasma exposures were 5.02 and 1.22 μM, respectively. One of the challenges associated with this series was to reduce P-gp mediated efflux in order to obtain acceptable brain exposure. This problem could be mitigated by introducing a lipophilic and electron-withdrawing chloro-substituent onto the pyridopyrazine core.

An alternative strategy by Schering/Merck was to prepare amide bioisosteres such as an oxadiazoline by forming a ring between the C-ring hydrogen-bond acceptor of E2012 and the right-hand methylene-linker, e.g. **23** [74]. Initial, monocyclic versions of this chemotype were found to be unstable under acidic conditions whereas bicyclic structures such as **23** had significantly improved stability. Furthermore, chemical stability was also dependent on the nature of the substituent at the quaternary centre. GSM **23** displayed an Aβ42 IC_{50} value of 29 nM in $HEK293^{Swe\text{-}Lon}$ cells

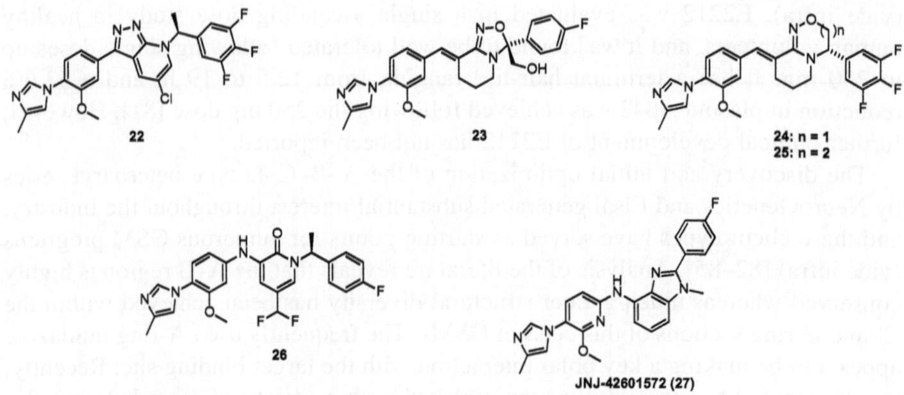

Fig. 7 Schering/Merck and Janssen GSMs

(*c*logP = 4.9), and this compares favourably to E2012, which had an Aβ42 IC_{50} of 58 nM under the same assay conditions. Given the excellent pharmacokinetic profile of compounds in the oxadiazoline series, this amide bioisostere was expanded further to afford the corresponding six-membered oxadiazine, e.g. **24** (Aβ42 IC_{50} = 33 nM, *c*logP = 6.6) [74, 89]. Immunoprecipitation/mass spectrometry analysis revealed that **23** and **24** reduce formation of Aβ42, Aβ40, and Aβ39 whereas production of the shorter peptides Aβ38 and Aβ37 is increased. Both lead compounds exhibited robust in vivo activity across multiple preclinical species. A 3 mg/kg oral dose of **24** in rat resulted in a 37% reduction of Aβ42 in cortex and a 35% reduction in CSF. In cynomolgous monkeys treated with **24** at 30 mg/kg p.o., 30% and 64% reductions of cortical and CSF Aβ42 were observed. This dose resulted in an exposure of ~6 μM of **24** in plasma and a brain/plasma ratio of about 1.4.

Inhibition of hERG remained a key challenge in the oxadiazoline and oxadiazine series. The two lead compounds **23** and **24** had hERG IC_{50} values of 3.3 and 1.0 μM, respectively, and additional heterocyclic cores were evaluated in an effort to address this problem. Ring expansion to generate fused oxadiazepines ultimately resulted in the discovery of **25** (Aβ42 IC_{50} = 39 nM, *c*logP = 7.1) [90]. This structural modification not only retained comparable in vitro Aβ42-lowering activity, but it also reduced hERG activity with **25** only exhibiting 33% inhibition of hERG at 10 μM. When dosed in rat at 10 mg/kg p.o., **25** achieved a 61% reduction of CSF Aβ42 at the 3-h time point. The corresponding plasma AUC exposure was 15,800 nM × h with a B/P of 3.2. It is also worth noting that **25** did not exhibit CYP inhibition, which has been a reoccurring problem in the A–B–C–D-type heteroaryl GSM series. GSM **25** was tested for inhibition of CYP3A4, 2D6, and 2C9 and was found to have IC_{50} values greater than 30 μM against these isoforms.

A characterizing feature of GSMs **20–25** is the rigid olefin linker that joins the B- and the C-rings ("olefin-linked heteroaryl series"). A key design objective for multiple organizations has been to identify a suitable replacement for the electron-rich olefin, and a common tactic has involved using an amine linker ("amine-linked heteroaryl series") [83]. This approach was pioneered by Schering/Merck as exemplified by **26** (Aβ42 IC_{50} = 44 nM, *c*logP = 4.6, Fig. 7). The team hypothesized that the NH-linker could form an intramolecular hydrogen bond with the C-ring pyridone carbonyl, thus maintaining the necessary rigidity [91]. The same design principles were utilized by researchers at Janssen, who further developed this concept to deliver GSMs with exquisite in vitro and in vivo Aβ42-lowering activity [92]. Janssen evaluated several heteroaromatic C-rings such as **27** that incorporate an sp^2 nitrogen as a hydrogen-bond acceptor in the same location relative to the NH-linker as the pyridone carbonyl of **26**. Imidazopyridines were initially explored as a pyridone bioisostere, and this was followed by benzimidazoles, indazoles, and oxazoles. Introduction of a methyl substituent on the benzimidazole C-ring was found to have a big impact on potency. Presumably, this substituent serves as a conformational control element to increase the dihedral angle between the C- and the D-rings thus locking the molecule in the putative pharmacologically active conformation. Furthermore, calculations indicated that introduction of the methyl substituent results in an increase in the hydrogen-bond acceptor strength of the sp^2 nitrogen.

Ultimately, benzimidazole JNJ-42601572 (**27**) was selected for further profiling (Aβ42 IC_{50} = 16 nM, *c*logP = 6.1, Fig. 7) [92]. A 10 mg/kg oral dose in mouse afforded a 51% reduction of brain Aβ42 at the 4-h time point. The corresponding brain and plasma exposures of **27** at this time point were 2.8 and 5.3 μM, respectively. A 36% reduction of brain Aβ42 was observed in rat at the same dose and time point with a concomitant increase in both Aβ37 and Aβ38 (402% and 49%, respectively) [92]. JNJ-42601572 was further evaluated in beagle dog at 20 mg/kg p.o., and this resulted in a 19% and 48% reduction of CSF Aβ42 at the 4- and 8-h time points, respectively. As with a majority of GSMs described in the literature up to this point, **27** maintained high aromaticity and high lipophilicity (*c*logP = 6.1), and this may explain the signs of liver toxicity that were observed in dogs when dosed at 20 mg/kg p.o.

Connection of the B- and the C-rings via an amine linker has been explored by a number of additional companies including Hoffmann-La Roche, AstraZeneca, BMS, Boehringer Ingelheim, and Forum. A frequently used strategy has been to incorporate an aminopyrimidine C-ring, and often to introduce a second, aliphatic ring as a conformational control element (Fig. 8). Hoffmann-La Roche has reported the discovery of GSMs such as RO-02 (**28**), which has an Aβ42 IC_{50} of 15 nM in the HEK293/sw cells, and 13 nM in rat primary cortical neurons (*c*logP = 5.1). This compound served as a starting point for the development of a biotinylated photoaffinity probe to study the mechanism of action of this class of GSMs [46]. A similar series of GSMs has been reported by AstraZeneca as highlighted by AZ4800 (**29**) and AZ3303 (**30**) [93]. These compounds were tested in HEK APPswe cells and exhibited Aβ42 IC_{50} values of 26 and 74 nM, respectively. As in the case heteroaryl GSMs described previously, the series from AstraZeneca increased formation of Aβ37 and Aβ38. It is interesting to note that while **29** exhibited a stronger increase in Aβ37 than Aβ38, the opposite trend was observed for **30** where a more robust increase was seen for Aβ-38. This finding suggests that structural modification even within the same series can impact the Aβ profile. In vivo efficacy of **29** and **30** was evaluated in wild-type mouse. At a dose of

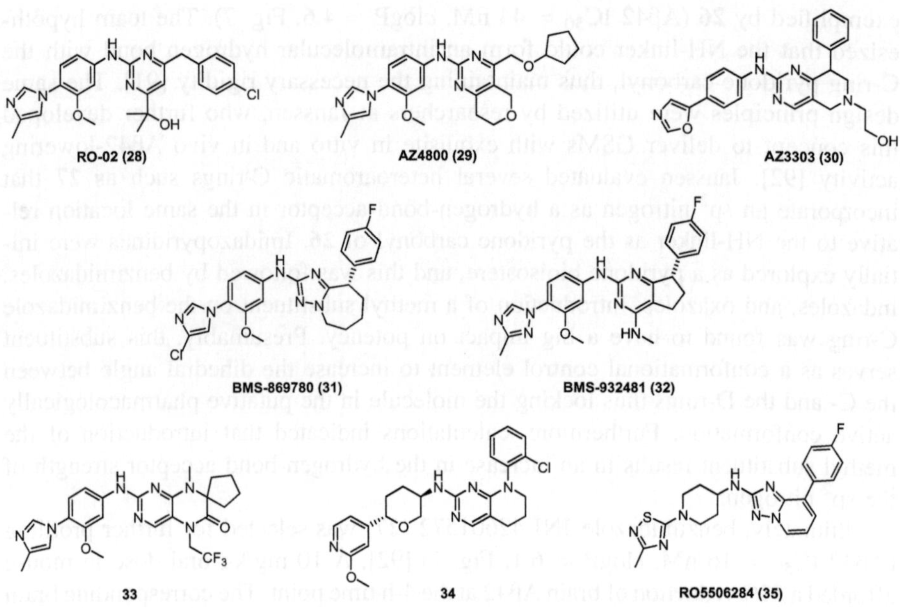

Fig. 8 Hoffmann-La Roche, AstraZeneca, BMS, Boehringer Ingelheim, and Forum GSMs

300 μmol/kg p.o. (130 mg/kg for **29** and 124 mg/kg for **30**), GSM **29** and **30** achieved 46% and 25% reduction of brain Aβ42, respectively. The corresponding free brain concentrations at these doses were 880 and 570 nM. GSM **29** is noteworthy in that it represents one of the first examples where robust Aβ42-lowering activity has been maintained while using a fully saturated ring system as the lipophilic D-ring. This resulted in a significant improvement in physicochemical properties such as reduced aromaticity, increased sp^3/sp^2 ratio (fsp^3), and reduced lipophilicity. GSM **29** has a *c*logP value of 3.3, which has generally been challenging to achieve – both within the NSAID-derived acid series as well as the non-NSAID heteroaryl series.

Starting from lead matter identified in a high-throughput screen, the team at BMS has developed potent GSMs that contain an aminotriazole C-ring. These efforts resulted in preclinical candidate BMS-869780 (**31**), which has an Aβ42 IC_{50} of 5.6 nM in H4-APPsw cells (*c*logP = 5.5) [94]. In vivo efficacy was extensively characterized in time-course studies in rats and triple transgenic mice to establish pharmacokinetic/pharmacodynamics (PK/PD) relationships and to enable human dose prediction. Through these studies and subsequent PK/PD modelling, in vivo IC_{50} values were determined for both species: The mouse and rat unbound plasma Aβ42 IC_{50} values were 22 and 5.1 nM, respectively, which correlates well with the in vitro IC_{50}. The PK profile of BMS-869780 was assessed in rat, dog, and non-human primate from which a projected human clearance of 5.6 mL/min/kg, half-life of 13 h, and oral bioavailability of 28% were established. A 25% reduction of brain Aβ42 human was projected to require a dose of 700 mg/day. Unfortunately, hepatotoxicity (lipidosis) was observed in a 4-day rat safety study at doses of 10 mg/kg/day that afforded a mean plasma C_{max} concentration of 1.9 μM. The toxicity was attributed to agonism of the pregnane X receptor (PXR), which is a nuclear hormone receptor involved in induction of cytochrome P450 3A4 (CYP3A4). Because of the high projected human dose and inadequate therapeutic index, further development of BMS-869780 was halted.

Continued efforts led to the identification of an alternative candidate molecule, BMS-932481 (**32**), where the chloroimidazole has been replaced with a methyltriazole and the central aminotriazole C-ring has been replaced with an aminopyrimidine [95]. This compound has similar potency and lipophilicity as the previous candidate (Aβ42 IC_{50} of 5.5 nM, *c*logP = 5.3). In vivo efficacy was established in rat at doses of 1, 5, and 10 mg/kg administered intravenously to minimize variability, and Aβ42 was maximally reduced in brain by about 75% at the 10 mg/kg dose. When administered orally in dog at 30 mg/kg, Aβ42 and Aβ40 were both maximally reduced by about 65% in CSF at the 24-h time point. Robust increases in Aβ37 and Aβ38 were observed (5-fold and 1.8-fold for Aβ37 and Aβ38, respectively), which is consistent with the profile of A–B–C–D-type heteroaryl GSMs. Finally, in vivo efficacy was also established in non-human primate, where an intravenous dose of 15 mg/kg afforded approximately a 75% reduction of CSF Aβ42 12–24 h post-dose.

In November 2012, BMS announced entering Phase I clinical trials with BMS-932481, and the clinical data has recently been published [95, 96]. BMS-932481 was initially evaluated in a single ascending dose study with oral doses ranging from 10 to 1,200 mg. The compound was generally well tolerated, and a 50% reduction of CSF Aβ42 was achieved in healthy adults following a single dose of 900 mg that afforded a plasma C_{max} exposure of 1.68 μM. BMS-932481 was subsequently evaluated in a multiple ascending dose study with doses of 50 and 100 mg for 28 days and 200 mg for 14 days. The corresponding reductions in CSF Aβ42 were 38%, 46%, and 69–77% for the three doses, respectively. A similar reduction of Aβ40 was achieved along with robust and dose-dependent increases in Aβ37 and Aβ38. The PK/PD relationship in human closely matched that obtained in preclinical species. Significant drug accumulation was observed in the multiple ascending dose study, and at the last day of dosing, C_{max} exposures reached 0.66, 1.50, and 4.43 μM for the three doses, respectively. Administration of BMS-932481 at a dose of 200 mg/day was halted after 14 days due to changes in liver enzymes (ALT). Ultimately, the development of BMS-932481 was terminated because the reduced therapeutic index due to liver toxicity would only enable reductions of CSF Aβ42 up to 25%.

Boehringer Ingelheim has also pursued *N*-linked heteroaryl GSMs and has published three patent applications describing compounds such as **33** that incorporate an aminopyrimidine C-ring [97–99]. It is interesting to note that this modulator achieved good in vitro activity (Aβ42 IC_{50} = 22 nM, *c*logP = 5.0) without an aromatic D-ring. Forum has recently published three patent applications that also describe GSMs incorporating an aminopyrimidine moiety as the C-ring hydrogen-bond acceptor [100–102]. Although many of the exemplified compounds utilize the typical arylimidazole A–B unit, certain compounds such as **34** (Aβ42 IC_{50} = 41 nM, *c*logP = 5.3) achieve good in vitro activity with a saturated ring system such as a tetrahydropyran. This represents a significant departure from a majority of the GSMs described it the patent literature and offers an alternative to the commonly used (and potentially problematic) aniline B-ring.

Roche has combined a thiadiazole and a piperidine ring as the as A–B moiety and introduced conformational constraint in the C-ring to afford GSMs with good Aβ-lowering activity [103]. GSM RO5506204 (**35**) has an Aβ42 IC_{50} of 25.7 nM in H4 neuroglioma cells overexpressing human APP-Swe and reduced lipophilicity (*c*logP = 3.4). Reductions of brain Aβ42 of 40%, 48%, and 73% were achieved at the 4-h time point when dosed in APP-Swe mouse at 10, 30, and 100 mg/kg p.o., respectively. The effect on plaque reduction was then evaluated in a chronic study in the TG2576 mouse model. GSM RO5506204 was dosed for 6 months at 30 mg/kg/day p.o., whereupon a 40% reduction of insoluble Aβ42 was observed relative to the vehicle group. Plaque deposition was also examined using positron emission tomography (PET) imaging and was found to show good correlation with terminal analysis of Aβ levels. Analysis of fibrillar Aβ plaques using methoxy-X04 staining indicated a 56% reduction relative to the vehicle group. Interestingly, there were considerably fewer small plaques in the GSM-treated animal group relative to the vehicle group, indicating that GSM treatment may have a significant effect on de novo plaque formation. The authors propose that a combination of a GSM with an Aβ antibody, which targets existing plaques, may have the potential for increased therapeutic benefit relative to either treatment alone.

In addition to Roche, other companies have also explored significant modifications to the A–B section of the A–B–C–D-type heteroaryl series. An interesting example has been published by researchers at Merck, who identified a novel chemotype through a high-throughput screening campaign. This series contains a methoxyphenyl piperazine in the A–B region, and a pyrimidine unit as the central hydrogen-bond acceptor [104]. Further optimization by replacing the pyrimidine with a purine moiety afforded GSM leads such as **36** (Fig. 9) [105]. This structural modification resulted in a tenfold boost in potency (Aβ42 IC_{50} = 21 nM in SH-SY5Y cells stably overexpressing the β-CTF SPA4CT, *c*logP = 8.0). In addition to the methoxyphenyl moiety, a number of additional groups were well tolerated including a 5-methyl-1,3-pyrimidine. Interestingly, **36** is about 16-fold selective for Aβ42 over Aβ40 in vitro, and this is borne out in vivo using the APP-YAC transgenic mouse model. When dosed at 100 mg/kg p.o., **36** selectively reduces brain Aβ42 by 73% at

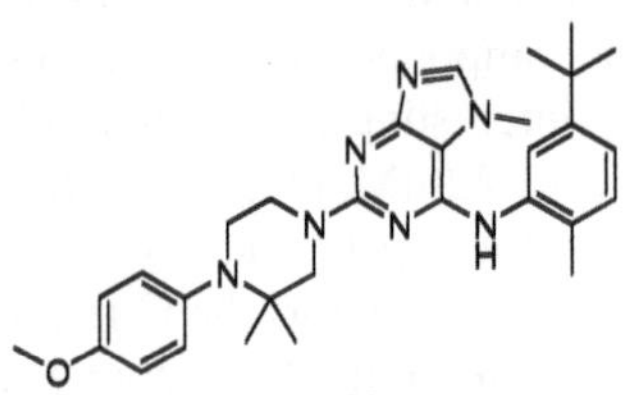

36

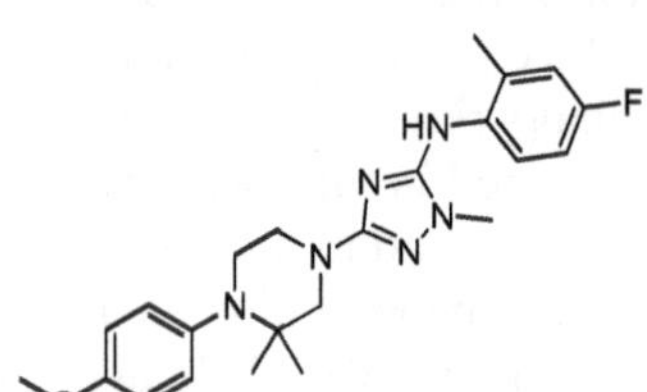

37

Fig. 9 Merck and Janssen GSMs

the 6-h time point. The corresponding plasma and brain exposure was 23 and 7.8 μM, respectively. More recently, Janssen has described a similar chemotype, where the purine of **36** has been replaced with a triazole, e.g. **37** [106]. Aβ42-lowering activity was measured in SKNBE2 cells (wtAPP 695), and **37** displayed good in vitro activity (Aβ42 IC_{50} = 50 nM, *c*logP = 5.5) [103]. This translated into a robust response in vivo with a 45% reduction of brain Aβ42 in rodents at a dose of 30 mg/kg p.o.

The strategy of directly connecting the B- and the C-rings has also been explored within the original arylimidazole moiety "*C*-linked heteroaryl series" [83]. GSK has utilized a pyridine and a pyridazine as the C-ring hydrogen-bond acceptor and has directly attached it to the B-ring to afford GSMs such as **38** (Fig. 10) [107–109]. A-β-lowering activity was evaluation in SHSY5Y-APPswe cells, and **38** displayed an IC_{50} value of 63 nM (*c*log*P* = 5.2). Combined with a good pharmacokinetic profile, this translated into robust, dose-dependent Aβ-lowering in vivo. When dosed in rat at 10, 30, and 100 mg/kg p.o., brain Aβ42 was reduced by ca 20%, 40%, and 75% for the three doses, respectively.

As discussed above, achieving robust Aβ-lowering activity (both in vitro and in vivo) while maintaining drug-like physicochemical properties has continued to be a significant challenge for the GSM field. In particular, good in vitro potency has often come at the expense of high lipophilicity, which in turn has reduced the odds of achieving adequate safety margins [110]. After encountering liver toxicity with **27**, the team at Janssen focused on reducing lipophilicity and aromaticity, resulting in compounds such as *C*-linked GSM **39** (Fig. 10) [111, 112]. This compound has an Aβ42 IC_{50} of 56 nM, but with dramatically reduced lipophilicity relative to **27** (*c*logP of 3.1 and 6.1 for **39** and **27**, respectively). When dosed orally at 20 mg/kg in dog, GSM **39** afforded a 30–40% maximal reduction of CSF Aβ42 at the 4- and 8-h time points, and there were no signs of liver toxicity in spite of high plasma exposure (24 μM). The safety profile of **39** was further evaluated in a 2-week safety study in rat at 200 mg/kg/day, and no liver toxicity was observed. Although this result represents a significant advancement for the program, the authors note that

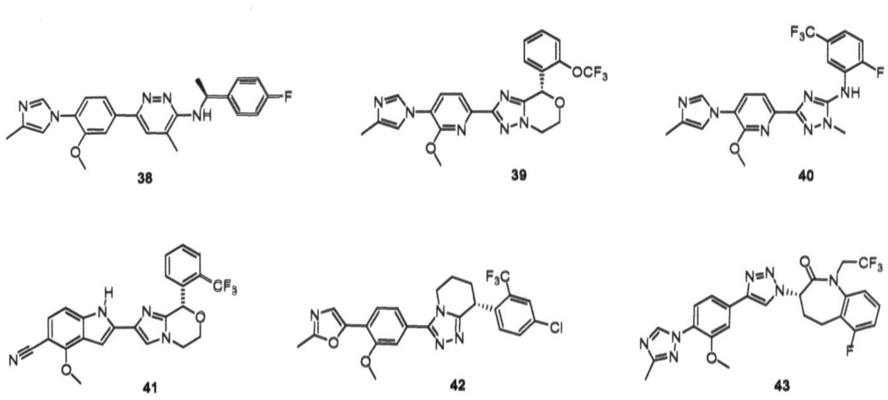

Fig. 10 GSK, Janssen, Takeda, and Merck GSMs

the high plasma exposure required for in vivo efficacy of **39** would not be suitable for chronic treatment, and additional gains in in vivo efficacy would be therefore be required.

Another approach used by Janssen to optimize the *C*-linked triazole GSMs involved using an aniline linker between the central triazole and the terminal phenyl ring along with an *N*-methyl substituent on the triazole as a conformational control element (e.g. **40**) [113]. Although **40** is more potent than **39** (Aβ42 $IC_{50} = 19$ nM), it is also considerably more lipophilic (*c*logP = 4.8). Aminotriazole **40** exhibited robust central efficacy in a time-course study in non-transgenic mouse. Brain Aβ42 was maximally reduced by 61% at the 4-h time point following a 30 mg/kg oral dose. Efficacy was also assessed in dog, where a dose of 20 mg/kg p.o. resulted in 56% reduction of CSF Aβ42 at the 8-h time point. No changes in liver enzymes (ALT and AST) were detected in a 1-week dog safety study at doses of 10 and 20 mg/kg/day, and only a small change in bilirubin was observed. Thus, **40** appears to have an improved safety profile relative to Janssen's early lead compound **27**.

Janssen has continued their search for new chemotypes with the goal of further improving alignment of potency and physicochemical properties. In early 2013, a patent application was published that disclosed compounds such as **41** that incorporate a novel A–B moiety [114]. In this series, the commonly used arylimidazole has been replaced with a cyanoindole unit. Compound **41** was reported to have an Aβ42 IC_{50} value of 20 nM and *c*logP of 3.9. Takeda has pursued chemical matter related to that described by Janssen [115]. Combination of an oxazolylphenyl A–B moiety with known C-ring heterocycles from the GSM literature afforded compounds such as **42** (Aβ42 IC_{50} of 60 nM in rat primary neuronal cells, *c*logP = 4.6, measured logD = 3.5). This compound reduced brain Aβ42 by 50% following a 10 mg/kg oral dose in mouse.

Scientists at Merck have designed GSMs that incorporate a 1,2,3-triazole linked to a benzoazepinone [116, 117]. Initial analogs in this series did not have adequate oral bioavailability in dog and monkey [116], but this problem could be addressed by modifications to the phenylimidazole A–B moiety leading to modulators such as **43** (Fig. 10) [117]. This compound has an Aβ42 IC_{50} of 42 nM in SH-SY5Y cells overexpressing APP C-terminal fragment SPA4CT (*c*logP = 3.8). Oral bioavailability in rat and dog was 25% and 84%, respectively, and brain Aβ42 was reduced by 38% when dosed at 50 mg/kg p.o. in the APP-YAC transgenic mouse model. Finally, robust central efficacy was also achieved in non-transgenic rat with an ED_{50} value of 4.7 mg/kg.

Pfizer has also pursued non-NSAID-derived GSMs using E2012 as a starting point [76]. However, while a majority of the previously highlighted chemotypes relied on a heterocyclic C-ring as the central hydrogen-bond acceptor, Pfizer retained the amide carbonyl of E2012 while removing the cinnamide olefin and directly attached the carbonyl to the A–B arylimidazole moiety. A conformationally restricted benzofuran unit was introduced as the D-ring with the goal of improving potency while moving toward improved physicochemical space. These efforts ultimately led to GSM **44** (Fig. 11), which has a *c*logP of 4.0 and moderate in vitro potency: Aβ42 $IC_{50} = 188$ nM in CHO-APP cells (E2012 has an Aβ42 IC_{50} value of

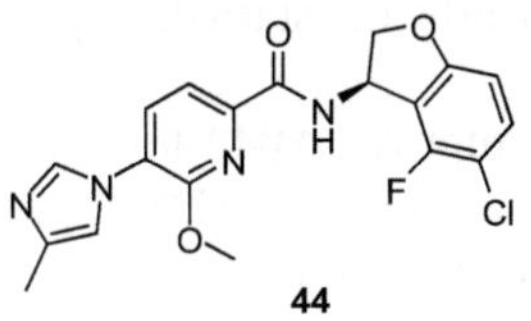

44

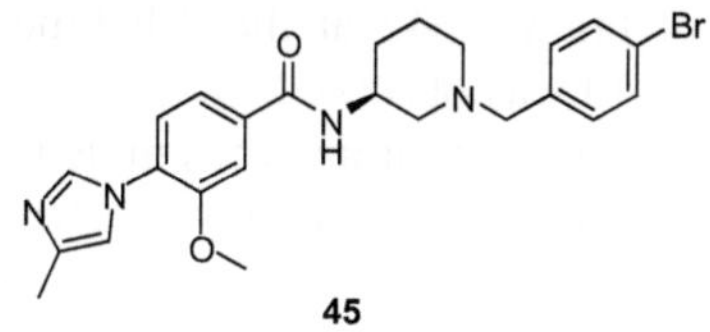

45

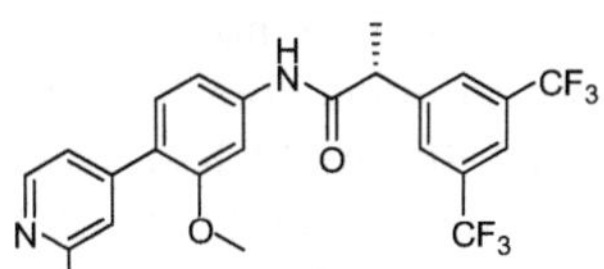

46

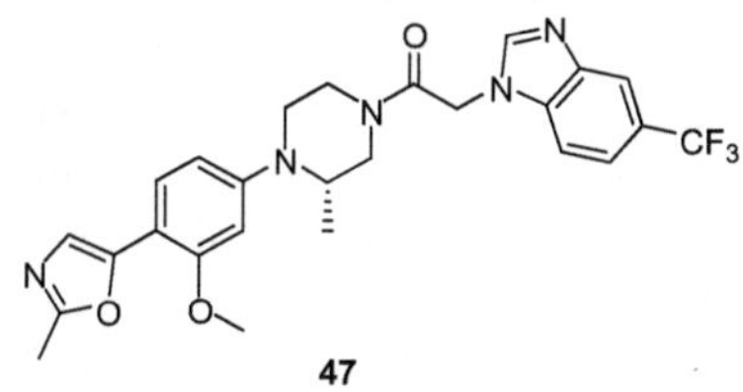

47

Fig. 11 Pfizer, Dainippon, Amgen, and Takeda GSMs

143 nM in the same assay). In vivo efficacy was assessed in guinea pig, and a 36% reduction of brain Aβ42 was achieved at the 4-h time point when dosing **44** orally at 100 mg/kg.

Related examples of GSMs that rely on a carbonyl as the central hydrogen-bond acceptor have recently been reported in the literature. Dainippon has disclosed amide GSMs such as **45** that incorporates a 3-aminopiperidine linker between the amide carbonyl and the terminal phenyl ring [118]. This compound has an Aβ42 IC_{50} of 250 nM in rat primary neurons (*c*logP = 5.0), and a 55% reduction of Aβ42 was observed in hippocampus when dossed in rat at a 100 mg/kg p.o. Amgen has reported GSMs with a reversed amide linker and a 2-methyl pyridine replacing the ubiquitous 4-methyl imidazole as exemplified by **46** (Aβ42 IC_{50} = 260 nM in HEK293 cell stably expressing the Swedish mutation, *c*logP = 5.3). This compound afforded a 33% reduction of Aβ42 in brain at the 4-h time point when dosed orally in rat at 100 mg/kg [119]. Although in vitro potency could be improved tenfold relative to **46**, this was achieved at the expense of significantly increased lipophilicity resulting in reduced permeability and poor in vivo efficacy. Finally, Takeda has reported amide GSMs such as **47** that incorporate a piperazine linker between the A–B-ring system and the carbonyl and where the imidazole has been replaced with an oxazole [120]. In vitro potency was determined in primary rat neuronal cells, and potency is reported as inflection points (IPs) along with maximum percent reduction of Aβ42, since many compounds did not reduce Aβ42 by 100%. The lead GSM **47** has an IP of 21 nM with a 60% maximum reduction of Aβ42 (*c*logP = 4.1). Interestingly, many compounds in this series did not reduce formation of Aβ40. In fact, **47** was profiled in IMR-32 human neuroblastoma cells where a *slight increase* in Aβ40 was observed along with robust reduction of Aβ42. This stands in sharp contrast to the majority of A–B–C–D heteroaryl GSMs which typically reduce both Aβ42 and Aβ40. The in vitro selectivity of **47** was recapitulated in vivo where a 100 mg/kg oral dose in C5BL/6 mice resulted in a 59%

reduction of brain Aβ42 at the 3-h time point while no statistically significant change in Aβ40 was observed.

Further efforts to optimize the amide GSMs to improve Aβ42-lowering activity while maintaining acceptable physicochemical properties and ADME profile have proven to be very challenging. To address this issue, the team at Pfizer shifted focus toward the A–B arylimidazole moiety with the goal of developing a novel, more polar heterocyclic core (Fig. 12). This was accomplished by introducing conformational constraint in the form of a two-carbon linker to connect amide NH to the pyridyl nitrogen and in the process converting the B-ring methoxy substituent to a pyridone carbonyl. The resultant pyridopyrazine-1,6-dione core (i.e. **48**) had significantly improved alignment of potency and ADME (Aβ42 IC_{50} = 103 nM in CHO-APP cells) as well as reduced lipophilicity (*c*logP = 3.3) [121, 122]. A dose-dependent reduction of brain Aβ42 was observed with reductions of 19%, 44%, and 68% at the 4-h time point when dosed in guinea pig at 10, 30, and 90 mg/kg p.o., respectively. The corresponding brain exposures at this time point and doses were 2.9, 9.2, and 52 μM indicating that further gains in vitro and in vivo efficacy would be needed. Through strategic use of fluorine to design lipophilic substituents with improved metabolic stability, this series was further optimized to deliver PF-06442609 (**49**), which exhibits excellent in vitro potency without a concomitant increase in lipophilicity (Aβ42 IC_{50} = 6 nM, *c*logP = 3.1) [123]. However, while excellent activity was achieved in vitro, this compound still required high drug exposure to achieve a robust reduction of brain Aβ42 in a guinea pig time-course experiment: A 41% reduction in brain Aβ42 was measured at the 3-h time point following a 60 mg/kg oral dose, and the corresponding total and unbound brain exposures were 17.3 μM and 225 nM, respectively.

During this time, naphthyl GSM **50** (Aβ42 IC_{50} = 42 nM, *c*logP = 3.3) was identified as a promising starting point for further optimization in part due to its favourable ligand efficiency (LE = 0.35). A combination of parallel medicinal chemistry and design of individual analogues led to the discovery of indoles as a suitable replacement for the naphthyl unit [124]. The resulting GSM **51** once again demonstrated that the pyridopyrazine-1,6-dione series had the potential to deliver

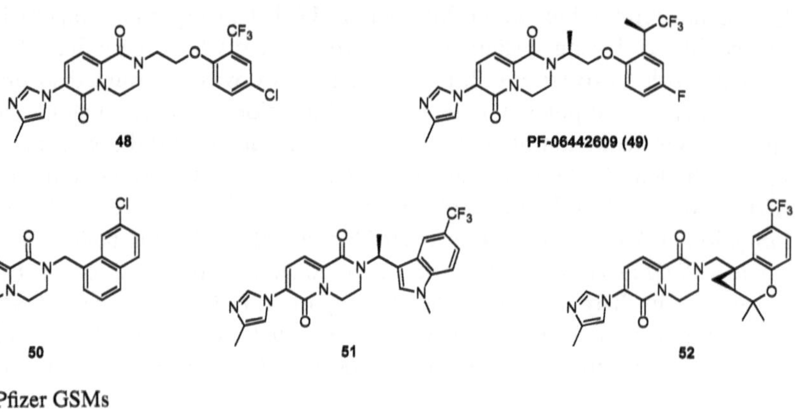

Fig. 12 Pfizer GSMs

GSMs with good alignment of in vitro potency, ADME, and acceptable lipophilicity: GSM **51** has an Aβ42 IC_{50} of 6 nM and *c*logP = 3.1. However, this compound was shown to be a substrate for P-glycoprotein (P-gp) efflux, which contributed to reduced central efficacy, and alternative heterocyclic replacements for the naphthyl unit were therefore required. Toward this end, the naphthyl moiety was replaced with a chromene, which was further refined into cyclopropyl chromane GSM **52** (Aβ42 IC_{50} = 4.9 nM, *c*logP = 3.2) [125]. This compound demonstrated dose-dependent and sustained reductions of brain Aβ42 at low unbound brain concentrations, thus overcoming significant hurdles encountered with earlier compounds. Reductions of 40% and 62% were observed at the 4-h time point following a 10 and 40 mg/kg oral dose, respectively, and the corresponding unbound brain concentrations were 11 and 67 nM. The successful identification of **52** was accomplished by using design tactics such as molecular modelling, designing conformational control elements to lock the compound in the putative bioactive conformation, blocking metabolic soft spots to increase LipMetE [126], and using the unpaced isolated heart Langendorff model to triage compounds with respect to cardiovascular safety. Pfizer has disclosed additional, related GSMs in two recent patent applications [127, 128]. Likewise, Janssen has pursued a similar approach and also arrived at the pyridopyrazine-1,6-dione heterocyclic core as disclosed in the patent literature [129–131]. Discovery of the pyridopyrazine-1,6-dione based GSMs and subsequent optimization to provide highly potent compounds in good physicochemical property space represents a key milestone for the GSM field.

In July 2016, Pfizer reported the first clinical data from a Phase 1 study with GSM clinical candidate PF-06648671 (structure not disclosed) [132]. The compound was evaluated in a double-blind, placebo-controlled single ascending dose study with doses ranging from 2 to 360 mg. The terminal half-life was found to be between 13.9 and 23.1 h, and a 56.9% reduction of plasma Aβ42 was reported following a single dose of 360 mg. GSM PF-06648671 was reported to be well tolerated at all doses, and therefore was advanced to a multiple ascending dose study to measure Aβ biomarkers in CSF at steady state.

2.3 *Natural Product-Derived γ-Secretase Modulators*

Satori Pharmaceuticals has reported a novel class of GSMs that originated from triterpene glycoside **53** (Aβ42 IC_{50} = 100 nM in H4 cells overexpressing human APP, Fig. 13) [133]. This compound was originally isolated from the black cohosh plant and was identified from a screen of a natural products collection. Interestingly; this class of GSMs reduces both Aβ42 and Aβ38 and increases Aβ37 and Aβ39 while Aβ40 remains relatively unchanged. The initial lead had poor metabolic stability, and lead optimization involved reduction of the olefin, replacement of the labile acetate with an ethyl substituent, and removal of the glycoside to install a substituted morpholine [134, 135]. These efforts ultimately resulted in preclinical candidate SPI-1865 (**54**), which has an Aβ42 IC_{50} of 106 nM in CHO-2B7 cells

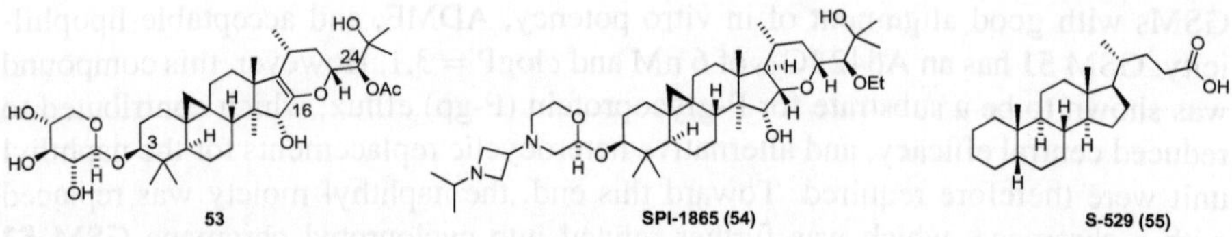

53 SPI-1865 (54) S-529 (55)

Fig. 13 Satori and University of Florida GSMs

expressing human wild-type APP [136, 137]. Dose-dependent reductions of brain Aβ42 were demonstrated in an acute rat study with doses of 10, 30, and 100 mg/kg p.o. resulting in brain Aβ42 reductions of 21%, 37%, and 50%, respectively, at 24 h post-dose. The corresponding brain exposures at this time point were 2.8, 11, and 33 μM, respectively. Development of SPI-1865 was eventually halted due to adverse effects on adrenal function in monkeys [85].

Jung and co-workers have reported on acidic steroids as another new class of GSMs [138]. Among the compounds studied, S-529 (**55**) was the most potent analogue with an Aβ42 IC_{50} value of 5.68 μM in CHO-2B7 cells. Unlike the steroidal GSMs from Satori, S-529 and related compounds display an Aβ modulatory profile that is similar to the NSAID-derived carboxylic acid class of GSMs, where Aβ42 is selectively reduced over Aβ40 with a concomitant increase in Aβ38. Closely related non-acidic steroids such as the corresponding esters act as inverse GSMs (iGSMs) resulting in an increase of Aβ42. The fact that these acidic steroids are derived from normal oxidative metabolism of cholesterol suggests that endogenous GSMs may exist.

3 Summary

GSMs reduce Aβ42 levels without affecting total Aβ by increasing shorter Aβ species such as Aβ38 and/or Aβ37. This generally results in a reduction in the Aβ42/Aβ40 ratio and an even larger reduction in the ratio of Aβ42 to shorter Aβ fragments including Aβ38 and Aβ37. This is particularly promising since Aβ42 and oligomers thereof are thought to represent major neurotoxic species contributing to AD [139, 140] and FAD presenilin mutations generally result in an increase in the Aβ42/Aβ40 ratio. The early clinical results with aducanumab, a monoclonal antibody that targets aggregated Aβ, showing a reduction of brain Aβ plaques and a slowing of clinical decline in patients with prodromal or mild AD are encouraging and provide support for the amyloid hypothesis [141]. The recent cryo-EM studies on the γ-secretase complex, taken together with development of improved GSMs that are progressing in the clinic, represent another major step forward toward testing the amyloid hypothesis. Ongoing and future studies to understand the GSM binding sites on γ-secretase and mechanism of action could reveal new insights into how γ-secretase contributes to AD and inform future design of γ-secretase targeted therapies.

References

1. Glenner GG, Wong CW (1984) Alzheimer's disease: initial report of the purification and characterization of a novel cerebrovascular amyloid protein. Biochem Biophys Res Commun 120:885–890
2. Wischik CM, Novak M, Thogersen HC, Edwards PC, Runswick MJ, Jakes R, Walker JE, Milstein C, Roth M, Klug A (1988) Isolation of a fragment of tau derived from the core of the paired helical filament of Alzheimer disease. Proc Natl Acad Sci U S A 85:4506–4510
3. Tanzi RE (2012) The genetics of Alzheimer disease. Cold Spring Harb Perspect Med 2: a006296
4. Karran E, Mercken M, De Strooper B (2011) The amyloid cascade hypothesis for Alzheimer's disease: an appraisal for the development of therapeutics. Nat Rev Drug Discov 10:698–712
5. Selkoe DJ, Hardy J (2016) The amyloid hypothesis of Alzheimer's disease at 25 years. EMBO Mol Med 8:595–608
6. Wolfe MS (2012) γ-Secretase inhibitors and modulators for Alzheimer's disease. J Neurochem 120:89–98
7. Crump CJ, Johnson DS, Li Y-M (2013) Development and mechanism of γ-secretase modulators for Alzheimer's disease. Biochemistry 52:3197–3216
8. De Strooper B, Vassar R, Golde T (2010) The secretases: enzymes with therapeutic potential in Alzheimer disease. Nat Rev Neurol 6:99–107
9. Takami M, Nagashima Y, Sano Y, Ishihara S, Morishima-Kawashima M, Funamoto S, Ihara Y (2009) gamma-Secretase: successive tripeptide and tetrapeptide release from the transmembrane domain of beta-carboxyl terminal fragment. J Neurosci 29:13042–13052
10. Olsson F, Schmidt S, Althoff V, Munter LM, Jin S, Rosqvist S, Lendahl U, Multhaup G, Lundkvist J (2014) Characterization of intermediate steps in amyloid beta (Abeta) production under near-native conditions. J Biol Chem 289:1540–1550
11. Matsumura N, Takami M, Okochi M, Wada-Kakuda S, Fujiwara H, Tagami S, Funamoto S, Ihara Y, Morishima-Kawashima M (2014) gamma-Secretase associated with lipid rafts: multiple interactive pathways in the stepwise processing of beta-carboxyl-terminal fragment. J Biol Chem 289:5109–5121
12. Ran Y, Cruz PE, Ladd TB, Fauq AH, Jung JI, Matthews J, Felsenstein KM, Golde TE (2014) gamma-Secretase processing and effects of gamma-secretase inhibitors and modulators on long Abeta peptides in cells. J Biol Chem 289:3276–3287
13. Iwatsubo T, Odaka A, Suzuki N, Mizusawa H, Nukina N, Ihara Y (1994) Visualization of A beta 42(43) and A beta 40 in senile plaques with end-specific A beta monoclonals: evidence that an initially deposited species is A beta 42(43). Neuron 13:45–53
14. Duff K, Eckman C, Zehr C, Yu X, Prada CM, Perez-tur J, Hutton M, Buee L, Harigaya Y, Yager D, Morgan D, Gordon MN, Holcomb L, Refolo L, Zenk B, Hardy J, Younkin S (1996) Increased amyloid-beta42(43) in brains of mice expressing mutant presenilin 1. Nature 383:710–713
15. Johnson-Wood K, Lee M, Motter R, Hu K, Gordon G, Barbour R, Khan K, Gordon M, Tan H, Games D, Lieberburg I, Schenk D, Seubert P, McConlogue L (1997) Amyloid precursor protein processing and A beta42 deposition in a transgenic mouse model of Alzheimer disease. Proc Natl Acad Sci U S A 94:1550–1555
16. Jarrett JT, Berger EP, Lansbury PT Jr (1993) The carboxy terminus of the beta amyloid protein is critical for the seeding of amyloid formation: implications for the pathogenesis of Alzheimer's disease. Biochemistry 32:4693–4697
17. Economou NJ, Giammona MJ, Do TD, Zheng X, Teplow DB, Buratto SK, Bowers MT (2016) Amyloid beta-protein assembly and Alzheimer's disease: dodecamers of Abeta42, but not of Abeta40, seed fibril formation. J Am Chem Soc 138:1772–1775
18. Kumar-Singh S, Theuns J, Van Broeck B, Pirici D, Vennekens K, Corsmit E, Cruts M, Dermaut B, Wang R, Van Broeckhoven C (2006) Mean age-of-onset of familial Alzheimer

disease caused by presenilin mutations correlates with both increased Abeta42 and decreased Abeta40. Hum Mutat 27:686–695
19. Yan Y, Wang C (2007) Abeta40 protects non-toxic Abeta42 monomer from aggregation. J Mol Biol 369:909–916
20. Murray MM, Bernstein SL, Nyugen V, Condron MM, Teplow DB, Bowers MT (2009) Amyloid beta protein: Abeta40 inhibits Abeta42 oligomerization. J Am Chem Soc 131:6316–6317
21. Jan A, Gokce O, Luthi-Carter R, Lashuel HA (2008) The ratio of monomeric to aggregated forms of Abeta40 and Abeta42 is an important determinant of amyloid-beta aggregation, fibrillogenesis, and toxicity. J Biol Chem 283:28176–28189
22. Kuperstein I, Broersen K, Benilova I, Rozenski J, Jonckheere W, Debulpaep M, Vandersteen A, Segers-Nolten I, Van Der Werf K, Subramaniam V, Braeken D, Callewaert G, Bartic C, D'Hooge R, Martins IC, Rousseau F, Schymkowitz J, De Strooper B (2010) Neurotoxicity of Alzheimer's disease Abeta peptides is induced by small changes in the Abeta42 to Abeta40 ratio. EMBO J 29:3408–3420
23. Pauwels K, Williams TL, Morris KL, Jonckheere W, Vandersteen A, Kelly G, Schymkowitz J, Rousseau F, Pastore A, Serpell LC, Broersen K (2012) Structural basis for increased toxicity of pathological abeta42:abeta40 ratios in Alzheimer disease. J Biol Chem 287:5650–5660
24. McGowan E, Pickford F, Kim J, Onstead L, Eriksen J, Yu C, Skipper L, Murphy MP, Beard J, Das P, Jansen K, Delucia M, Lin WL, Dolios G, Wang R, Eckman CB, Dickson DW, Hutton M, Hardy J, Golde T (2005) Abeta42 is essential for parenchymal and vascular amyloid deposition in mice. Neuron 47:191–199
25. Hu X, Li X, Zhao M, Gottesdiener A, Luo W, Paul S (2014) Tau pathogenesis is promoted by Abeta1-42 but not Abeta1-40. Mol Neurodegener 9:52
26. De Strooper B, Iwatsubo T, Wolfe MS (2012) Presenilins and gamma-secretase: structure, function, and role in Alzheimer disease. Cold Spring Harb Perspect Med 2:a006304
27. Bai XC, Yan C, Yang G, Lu P, Ma D, Sun L, Zhou R, Scheres SH, Shi Y (2015) An atomic structure of human gamma-secretase. Nature 525:212–217
28. Bai XC, Rajendra E, Yang G, Shi Y, Scheres SH (2015) Sampling the conformational space of the catalytic subunit of human gamma-secretase. Elife 4:e11182
29. Henley DB, Sundell KL, Sethuraman G, Dowsett SA, May PC (2014) Safety profile of semagacestat, a gamma-secretase inhibitor: IDENTITY trial findings. Curr Med Res Opin 30:2021–2032
30. Coric V, van Dyck CH, Salloway S, Andreasen N, Brody M, Richter RW, Soininen H, Thein S, Shiovitz T, Pilcher G, Colby S, Rollin L, Dockens R, Pachai C, Portelius E, Andreasson U, Blennow K, Soares H, Albright C, Feldman HH, Berman RM (2012) Safety and tolerability of the gamma-secretase inhibitor avagacestat in a phase 2 study of mild to moderate Alzheimer disease. Arch Neurol 69:1430–1440
31. De Strooper B (2014) Lessons from a failed γ-secretase Alzheimer trial. Cell 159:721–726
32. Haapasalo A, Kovacs DM (2011) The many substrates of presenilin/gamma-secretase. J Alzheimers Dis 25:3–28
33. Crump CJ, Castro SV, Wang F, Pozdnyakov N, Ballard TE, Sisodia SS, Bales KR, Johnson DS, Li YM (2012) BMS-708,163 targets presenilin and lacks notch-sparing activity. Biochemistry 51:7209–7211
34. Mitani Y, Yarimizu J, Saita K, Uchino H, Akashiba H, Shitaka Y, Ni K, Matsuoka N (2012) Differential effects between γ-secretase inhibitors and modulators on cognitive function in amyloid precursor protein-transgenic and nontransgenic mice. J Neurosci 32:2037–2050
35. Weggen S, Eriksen JL, Das P, Sagi SA, Wang R, Pietrzik CU, Findlay KA, Smith TE, Murphy MP, Bulter T, Kang DE, Marquez-Sterling N, Golde TE, Koo EH (2001) A subset of NSAIDs lower amyloidogenic Abeta42 independently of cyclooxygenase activity. Nature 414:212–216
36. Sagi SA, Weggen S, Eriksen J, Golde TE, Koo EH (2003) The non-cyclooxygenase targets of non-steroidal anti-inflammatory drugs, lipoxygenases, peroxisome proliferator-activated

receptor, inhibitor of kappa B kinase, and NF kappa B, do not reduce amyloid beta 42 production. J Biol Chem 278:31825–31830
37. Eriksen JL, Sagi SA, Smith TE, Weggen S, Das P, McLendon DC, Ozols VV, Jessing KW, Zavitz KH, Koo EH, Golde TE (2003) NSAIDs and enantiomers of flurbiprofen target gamma-secretase and lower Abeta 42 in vivo. J Clin Invest 112:440–449
38. Green RC, Schneider LS, Amato DA, Beelen AP, Wilcock G, Swabb EA, Zavitz KH (2009) Effect of tarenflurbil on cognitive decline and activities of daily living in patients with mild Alzheimer disease: a randomized controlled trial. JAMA 302:2557–2564
39. Wilcock GK, Black SE, Hendrix SB, Zavitz KH, Swabb EA, Laughlin MA, Tarenflurbil Phase II Study investigators (2008) Efficacy and safety of tarenflurbil in mild to moderate Alzheimer's disease: a randomised phase II trial. Lancet Neurol 7:483–493
40. Chavez-Gutierrez L, Bammens L, Benilova I, Vandersteen A, Benurwar M, Borgers M, Lismont S, Zhou L, Van Cleynenbreugel S, Esselmann H, Wiltfang J, Serneels L, Karran E, Gijsen H, Schymkowitz J, Rousseau F, Broersen K, De Strooper B (2012) The mechanism of gamma-secretase dysfunction in familial Alzheimer disease. EMBO J 31:2261–2274
41. Chau DM, Crump CJ, Villa JC, Scheinberg DA, Li YM (2012) Familial Alzheimer disease presenilin-1 mutations alter the active site conformation of gamma-secretase. J Biol Chem 287:17288–17296
42. Okochi M, Tagami S, Yanagida K, Takami M, Kodama TS, Mori K, Nakayama T, Ihara Y, Takeda M (2013) gamma-Secretase modulators and presenilin 1 mutants act differently on presenilin/gamma-secretase function to cleave Abeta42 and Abeta43. Cell Rep 3:42–51
43. Fernandez MA, Klutkowski JA, Freret T, Wolfe MS (2014) Alzheimer presenilin-1 mutations dramatically reduce trimming of long amyloid beta-peptides (Abeta) by gamma-secretase to increase 42-to-40-residue Abeta. J Biol Chem 289:31043–31052
44. Kukar TL, Ladd TB, Bann MA, Fraering PC, Narlawar R, Maharvi GM, Healy B, Chapman R, Welzel AT, Price RW, Moore B, Rangachari V, Cusack B, Eriksen J, Jansen-West K, Verbeeck C, Yager D, Eckman C, Ye W, Sagi S, Cottrell BA, Torpey J, Rosenberry TL, Fauq A, Wolfe MS, Schmidt B, Walsh DM, Koo EH, Golde TE (2008) Substrate-targeting gamma-secretase modulators. Nature 453:925–929
45. Uemura K, Farner KC, Hashimoto T, Nasser-Ghodsi N, Wolfe MS, Koo EH, Hyman BT, Berezovska O (2010) Substrate docking to gamma-secretase allows access of gamma-secretase modulators to an allosteric site. Nat Commun 1:130
46. Ebke A, Luebbers T, Fukumori A, Shirotani K, Haass C, Baumann K, Steiner H (2011) Novel gamma-secretase enzyme modulators directly target presenilin protein. J Biol Chem 286:37181–37186
47. Ohki Y, Higo T, Uemura K, Shimada N, Osawa S, Berezovska O, Yokoshima S, Fukuyama T, Tomita T, Iwatsubo T (2011) Phenylpiperidine-type gamma-secretase modulators target the transmembrane domain 1 of presenilin 1. EMBO J 30:4815–4824
48. Crump CJ, Fish BA, Castro SV, Chau D-M, Gertsik N, Ahn K, Stiff C, Pozdnyakov N, Bales KR, Johnson DS, Li Y-M (2011) Piperidine acetic acid based γ-secretase modulators directly bind to presenilin-1. ACS Chem Neurosci 2:705–710
49. Crump CJ, Johnson DS, Li YM (2011) Target of gamma-secretase modulators, presenilin marks the spot. EMBO J 30:4696–4698
50. Jumpertz T, Rennhack A, Ness J, Baches S, Pietrzik CU, Bulic B, Weggen S (2012) Presenilin is the molecular target of acidic gamma-secretase modulators in living cells. PLoS One 7:e30484
51. Pozdnyakov N, Murrey HE, Crump CJ, Pettersson M, Ballard TE, Am Ende CW, Ahn K, Li YM, Bales KR, Johnson DS (2013) gamma-Secretase modulator (GSM) photoaffinity probes reveal distinct allosteric binding sites on presenilin. J Biol Chem 288:9710–9720
52. Takeo K, Tanimura S, Shinoda T, Osawa S, Zahariev IK, Takegami N, Ishizuka-Katsura Y, Shinya N, Takagi-Niidome S, Tominaga A, Ohsawa N, Kimura-Someya T, Shirouzu M, Yokoshima S, Yokoyama S, Fukuyama T, Tomita T, Iwatsubo T (2014) Allosteric regulation

of gamma-secretase activity by a phenylimidazole-type gamma-secretase modulator. Proc Natl Acad Sci U S A 111:10544–10549
53. Peretto I, Radaelli S, Parini C, Zandi M, Raveglia LF, Dondio G, Fontanella L, Misiano P, Bigogno C, Rizzi A, Riccardi B, Biscaioli M, Marchetti S, Puccini P, Catinella S, Rondelli I, Cenacchi V, Bolzoni PT, Caruso P, Villetti G, Facchinetti F, Del Giudice E, Moretto N, Imbimbo BP (2005) Synthesis and biological activity of flurbiprofen analogues as selective inhibitors of beta-amyloid(1)(-)(42) secretion. J Med Chem 48:5705–5720
54. Imbimbo BP, Frigerio E, Breda M, Fiorentini F, Fernandez M, Sivilia S, Giardino L, Calza L, Norris D, Casula D, Shenouda M (2013) Pharmacokinetics and pharmacodynamics of CHF5074 after short-term administration in healthy subjects. Alzheimer Dis Assoc Disord 27:278–286
55. Rogers K, Felsenstein KM, Hrdlicka L, Tu Z, Albayya F, Lee W, Hopp S, Miller MJ, Spaulding D, Yang Z, Hodgdon H, Nolan S, Wen M, Costa D, Blain JF, Freeman E, De Strooper B, Vulsteke V, Scrocchi L, Zetterberg H, Portelius E, Hutter-Paier B, Havas D, Ahlijanian M, Flood D, Leventhal L, Shapiro G, Patzke H, Chesworth R, Koenig G (2012) Modulation of gamma-secretase by EVP-0015962 reduces amyloid deposition and behavioral deficits in Tg2576 mice. Mol Neurodegener 7:61
56. Van BB, Chen JM, Treton G, Desmidt M, Hopf C, Ramsden N, Karran E, Mercken M, Rowley A (2011) Chronic treatment with a novel γ-secretase modulator, JNJ-40418677, inhibits amyloid plaque formation in a mouse model of Alzheimer's disease. Br J Pharmacol 163:375–389
57. Hieke M, Ness J, Steri R, Dittrich M, Greiner C, Werz O, Baumann K, Schubert-Zsilavecz M, Weggen S, Zettl H (2010) Design, synthesis, and biological evaluation of a novel class of gamma-secretase modulators with PPARgamma activity. J Med Chem 53:4691–4700
58. Peng H, Talreja T, Xin Z, Cuervo JH, Kumaravel G, Humora MJ, Xu L, Rohde E, Gan L, Jung M-Y, Shackett MN, Chollate S, Dunah AW, Snodgrass-Belt PA, Arnold HM, Taveras AG, Rhodes KJ, Scannevin RH (2011) Discovery of BIIB042, a potent, selective, and orally bioavailable γ-secretase modulator. ACS Med Chem Lett 2:786–791
59. Page RM, Baumann K, Tomioka M, Perez-Revuelta BI, Fukumori A, Jacobsen H, Flohr A, Luebbers T, Ozmen L, Steiner H, Haass C (2008) Generation of Abeta38 and Abeta42 is independently and differentially affected by familial Alzheimer disease-associated presenilin mutations and gamma-secretase modulation. J Biol Chem 283:677–683
60. Page RM, Gutsmiedl A, Fukumori A, Winkler E, Haass C, Steiner H (2010) Beta-amyloid precursor protein mutants respond to gamma-secretase modulators. J Biol Chem 285:17798–17810
61. Hall A, Elliott RL, Giblin GM, Hussain I, Musgrave J, Naylor A, Sasse R, Smith B (2010) Piperidine-derived gamma-secretase modulators. Bioorg Med Chem Lett 20:1306–1311
62. Li T, Huang Y, Jin S, Ye L, Rong N, Yang X, Ding Y, Cheng Z, Zhang J, Wan Z, Harrison DC, Hussain I, Hall A, Lee DH, Lau LF, Matsuoka Y (2012) Gamma-secretase modulators do not induce Abeta-rebound and accumulation of beta-C-terminal fragment. J Neurochem 121:277–286
63. Mitani Y, Yarimizu J, Akashiba H, Shitaka Y, Ni K, Matsuoka N (2013) Amelioration of cognitive deficits in plaque-bearing Alzheimer's disease model mice through selective reduction of nascent soluble Abeta42 without affecting other Abeta pools. J Neurochem 125:465–472
64. Stanton MG, Hubbs J, Sloman D, Hamblett C, Andrade P, Angagaw M, Bi G, Black RM, Crispino J, Cruz JC, Fan E, Farris G, Hughes BL, Kenific CM, Middleton RE, Nikov G, Sajonz P, Shah S, Shomer N, Szewczak AA, Tanga F, Tudge MT, Shearman M, Munoz B (2010) Fluorinated piperidine acetic acids as gamma-secretase modulators. Bioorg Med Chem Lett 20:755–758
65. Baillie TA (2008) Metabolism and toxicity of drugs. Two decades of progress in industrial drug metabolism. Chem Res Toxicol 21:129–137
66. Am Ende CW, Fish BA, Johnson DS, Lira R, O'Donnell CJ, Pettersson MY, Stiff CM (2011) Aminocyclohexanes and aminotetrahydropyrans as γ-secretase modulators and their preparation and use for the treatment of neurological and psychiatric diseases. WO2011092611A1

67. Mitani Y, Akashiba H, Saita K, Yarimizu J, Uchino H, Okabe M, Asai M, Yamasaki S, Nozawa T, Ishikawa N, Shitaka Y, Ni K, Matsuoka N (2014) Pharmacological characterization of the novel gamma-secretase modulator AS2715348, a potential therapy for Alzheimer's disease, in rodents and nonhuman primates. Neuropharmacology 79:412–419
68. Cheng S, Comer DD, Mao L, Balow GP, Pleynet D (2004) Aryl compounds and uses in modulating amyloid β. WO2004110350A2
69. Kimura T, Kawano K, Doi E, Kitazawa N, Shin K, Miyagawa T, Kaneko T, Ito K, Takaishi M, Sasaki T, Hagiwara H (2005) Preparation of cinnamide, 3-benzylidenepiperidin-2-one, phenylpropynamide compounds as amyloid β production inhibitors. WO2005115990A1
70. Kounnas MZ, Danks AM, Cheng S, Tyree C, Ackerman E, Zhang X, Ahn K, Nguyen P, Comer D, Mao L, Yu C, Pleynet D, Digregorio PJ, Velicelebi G, Stauderman KA, Comer WT, Mobley WC, Li YM, Sisodia SS, Tanzi RE, Wagner SL (2010) Modulation of gamma-secretase reduces beta-amyloid deposition in a transgenic mouse model of Alzheimer's disease. Neuron 67:769–780
71. Values for *c*logP were calculated using the BIOBYTE (www.biobyte.com) program *c*logP, v.4.3
72. Wagner SL, Zhang C, Cheng S, Nguyen P, Zhang X, Rynearson KD, Wang R, Li Y, Sisodia SS, Mobley WC, Tanzi RE (2014) Soluble gamma-secretase modulators selectively inhibit the production of the 42-amino acid amyloid beta peptide variant and augment the production of multiple carboxy-truncated amyloid beta species. Biochemistry 53:702–713
73. ClinicalTrials.gov. https://clinicaltrials.gov/ct2/results?term=NGP-555&Search=Search. Accessed 4 Sept 2016
74. Sun Z-Y, Asberom T, Bara T, Bennett C, Burnett D, Chu I, Clader J, Cohen-Williams M, Cole D, Czarniecki M, Durkin J, Gallo G, Greenlee W, Josien H, Huang X, Hyde L, Jones N, Kazakevich I, Li H, Liu X, Lee J, MacCoss M, Mandal MB, McCracken T, Nomeir A, Mazzola R, Palani A, Parker EM, Pissarnitski DA, Qin J, Song L, Terracina G, Vicarel M, Voigt J, Xu R, Zhang L, Zhang Q, Zhao Z, Zhu X, Zhu Z (2012) Cyclic hydroxyamidines as amide isosteres: discovery of oxadiazolines and oxadiazines as potent and highly efficacious γ-secretase modulators in vivo. J Med Chem 55:489–502
75. Hashimoto T, Ishibashi A, Hagiwara H, Murata Y, Takenaka O, Miyagawa T (2010) E2012: a novel gamma-secretase modulator – pharmacology. Alzheimers Dement 6(Suppl):S242
76. Pettersson M, Johnson DS, Subramanyam C, Bales KR, am Ende CW, Fish BA, Green ME, Kauffman GW, Lira R, Mullins PB, Navaratnam T, Sakya SM, Stiff CM, Tran TP, Vetelino BC, Xie L, Zhang L, Pustilnik LR, Wood KM, O'Donnell CJ (2012) Design and synthesis of dihydrobenzofuran amides as orally bioavailable, centrally active γ-secretase modulators. Bioorg Med Chem Lett 22:2906–2911
77. Portelius E, Van Broeck B, Andreasson U, Gustavsson MK, Mercken M, Zetterberg H, Borghys H, Blennow K (2010) Acute effect on the Abeta isoform pattern in CSF in response to gamma-secretase modulator and inhibitor treatment in dogs. J Alzheimers Dis 21:1005–1012
78. Lu Y, Riddell D, Hajos-Korcsok E, Bales K, Wood KM, Nolan CE, Robshaw AE, Zhang L, Leung L, Becker SL, Tseng E, Barricklow J, Miller EH, Osgood S, O'Neill BT, Brodney MA, Johnson DS, Pettersson M (2012) Cerebrospinal fluid amyloid-beta (Abeta) as an effect biomarker for brain Abeta lowering verified by quantitative preclinical analyses. J Pharmacol Exp Ther 342:366–375
79. Nagy C, Schuck E, Ishibashi A, Nakatani Y, Rege B, Logovinsky V (2010) E2012, a novel γ-secretase modulator, decreases plasma amyloid-β (Aβ) levels in humans. Alzheimers Dement 6:S574
80. Nakano-Ito K, Fujikawa Y, Hihara T, Shinjo H, Kotani S, Suganuma A, Aoki T, Tsukidate K (2014) E2012-induced cataract and its predictive biomarkers. Toxicol Sci 137:249–258
81. Yu Y, Logovinsky V, Schuck E, Kaplow J, Chang MK, Miyagawa T, Wong N, Ferry J (2014) Safety, tolerability, pharmacokinetics, and pharmacodynamics of the novel gamma-secretase modulator, E2212, in healthy human subjects. J Clin Pharmacol 54:528–536

82. Pettersson M, Kauffman GW, am Ende CW, Patel NC, Stiff C, Tran TP, Johnson DS (2011) Novel γ-secretase modulators: a review of patents from 2008 to 2010. Expert Opin Ther Pat 21:205–226
83. Pettersson M, Stepan AF, Kauffman GW, Johnson DS (2013) Novel γ-secretase modulators for the treatment of Alzheimer's disease: a review focusing on patents from 2010 to 2012. Expert Opin Ther Pat 23:1349–1366
84. Oehlrich D, Berthelot DJC, Gijsen HJM (2011) γ-Secretase modulators as potential disease modifying anti-Alzheimer's drugs. J Med Chem 54:669–698
85. Hall A, Patel TR (2014) gamma-Secretase modulators: current status and future directions. Prog Med Chem 53:101–145
86. Caldwell JP, Bennett CE, McCracken TM, Mazzola RD, Bara T, Buevich A, Burnett DA, Chu I, Cohen-Williams M, Josein H, Hyde L, Lee J, McKittrick B, Song L, Terracina G, Voigt J, Zhang L, Zhu Z (2010) Iminoheterocycles as γ-secretase modulators. Bioorg Med Chem Lett 20:5380–5384
87. Qin J, Dhondi P, Huang X, Mandal M, Zhao Z, Pissarnitski D, Zhou W, Aslanian R, Zhu Z, Greenlee W, Clader J, Zhang L, Cohen-Williams M, Jones N, Hyde L, Palani A (2011) Discovery of fused 5,6-bicyclic heterocycles as γ-secretase modulators. Bioorg Med Chem Lett 21:664–669
88. Qin J, Zhou W, Huang X, Dhondi P, Palani A, Aslanian R, Zhu Z, Greenlee W, Cohen-Williams M, Jones N, Hyde L, Zhang L (2011) Discovery of a potent pyrazolopyridine series of γ-secretase modulators. ACS Med Chem Lett 2:471–476
89. Huang X, Zhou W, Liu X, Li H, Sun G, Mandal M, Vicarel M, Zhu X, Bennett C, McCraken T, Pissarnitski D, Zhao Z, Cole D, Gallo G, Zhu Z, Palani A, Aslanian R, Clader J, Czarniecki M, Greenlee W, Burnett D, Cohen-Williams M, Hyde L, Song L, Zhang L, Chu I, Buevich A (2012) Synthesis and SAR studies of fused oxadiazines as γ-secretase modulators for treatment of Alzheimer's disease. ACS Med Chem Lett 3:931–935
90. Li H, Qin J, Dhondi P, Zhou W, Vicarel M, Bara T, Cole D, Josien H, Pissarnitski D, Zhu Z, Palani A, Aslanian R, Clader J, Czarniecki M, Greenlee W, Cohen-Williams M, Hyde L, Song L, Zhang L, Chu I, Huang X (2013) The discovery of fused oxadiazepines as gamma secretase modulators for treatment of Alzheimer's disease. Bioorg Med Chem Lett 23:466–471
91. Huang X, Aslanian R, Zhou W, Zhu X, Qin J, Greenlee W, Zhu Z, Zhang L, Hyde L, Chu I, Cohen-Williams M, Palani A (2010) The discovery of pyridone and pyridazone heterocycles as γ-secretase modulators. ACS Med Chem Lett 1:184–187
92. Bischoff F, Berthelot D, De CM, MacDonald G, Minne G, Oehlrich D, Pieters S, Surkyn M, Trabanco AA, Tresadern G, Van BS, Velter I, Zaja M, Borghys H, Masungi C, Mercken M, Gijsen HJM (2012) Design and synthesis of a novel series of bicyclic heterocycles as potent γ-secretase modulators. J Med Chem 55:9089–9106
93. Borgegard T, Jureus A, Olsson F, Rosqvist S, Sabirsh A, Rotticci D, Paulsen K, Klintenberg R, Yan H, Waldman M, Stromberg K, Nord J, Johansson J, Regner A, Parpal S, Malinowsky D, Radesater AC, Li T, Singh R, Eriksson H, Lundkvist J (2012) First and second generation gamma-secretase modulators (GSMs) modulate amyloid-beta (Abeta) peptide production through different mechanisms. J Biol Chem 287:11810–11819
94. Toyn JH, Thompson LA, Lentz KA, Meredith JE Jr, Burton CR, Sankaranararyanan S, Guss V, Hall T, Iben LG, Krause CM, Krause R, Lin XA, Pierdomenico M, Polson C, Robertson AS, Denton RR, Grace JE, Morrison J, Raybon J, Zhuo X, Snow K, Padmanabha R, Agler M, Esposito K, Harden D, Prack M, Varma S, Wong V, Zhu Y, Zvyaga T, Gerritz S, Marcin LR, Higgins MA, Shi J, Wei C, Cantone JL, Drexler DM, Macor JE, Olson RE, Ahlijanian MK, Albright CF (2014) Identification and preclinical pharmacology of the gamma-secretase modulator BMS-869780. Int J Alzheimers Dis 2014:431858
95. Toyn JH, Boy KM, Raybon J, Meredith JE Jr, Robertson AS, Guss V, Hoque N, Sweeney F, Zhuo X, Clarke W, Snow K, Denton RR, Zuev D, Thompson LA, Morrison J, Grace J, Berisha F, Furlong M, Wang JS, Lentz KA, Padmanabha R, Cook L, Wei C, Drexler DM, Macor JE, Albright CF, Gasior M, Olson RE, Hong Q, Soares HD, AbuTarif M, Ahlijanian MK (2016)

Robust translation of gamma-secretase modulator pharmacology across preclinical species and human subjects. J Pharmacol Exp Ther 358:125–137

96. Soares HD, Gasior M, Toyn JH, Wang JS, Hong Q, Berisha F, Furlong MT, Raybon J, Lentz KA, Sweeney F, Zheng N, Akinsanya B, Berman RM, Thompson LA, Olson RE, Morrison J, Drexler DM, Macor JE, Albright CF, Ahlijanian MK, AbuTarif M (2016) The gamma-secretase modulator, BMS-932481, modulates Abeta peptides in the plasma and cerebrospinal fluid of healthy volunteers. J Pharmacol Exp Ther 358:138–150
97. Eickmeier C, Gerlach K, Heine N, Weber A, Gross U (2013) Preparation of spiroheterocycl-dihydropteridinones as modulators of γ-secretase for treating Alzheimer's disease. US20130225593A1
98. Gerlach K, Eickmeier C, Heine C, Heine N, Weber A (2013) Preparation of as dihydropteridinones as modulators of γ-secretase for treating Alzheimer's disease. US20130225549A1
99. Gerlach K, Heine N, Hobson S, Hoenke C, Weber A (2013) Preparation of dihydro-thienopyrimidines γ-secretase modulators. WO2013020992A1
100. Burnett DA, Bursavich MG, McRiner AJ (2015) Preparation of fused morpholinopyrimidines for treating neurodegenerative diseases. WO2015066696A1
101. Burnett DA, Bursavich MG, McRiner AJ (2015) Fused morpholinopyrimidines as beta-amyloid inhibitors and their preparation. WO2015109109A1
102. Burnett DA, Bursavich MG, McRiner AJ (2015) Preparation of fused morpholinopyrimidines for treating neurodegenerative diseases. WO2015066697A1
103. Brende M, Jaworska A, Herms J, Trambauer J, Rötzer C, Gildehaus F-J, Carlsen J, Cumming P, Bylund J, Luebbers T, Bartenstein P, Steiner H, Haass C, Baumann K, Rominger A (2015) Amyloid-PET predicts inhibition of de novo plaque formation upon chronic γ-secretase modulator treatment. Mol Psychiatry 20:1179–1187
104. Rivkin A, Ahearn SP, Chichetti SM, Kim YR, Li C, Rosenau A, Kattar SD, Jung J, Shah S, Hughes BL, Crispino JL, Middleton RE, Szewczak AA, Munoz B, Shearman MS (2010) Piperazinyl pyrimidine derivatives as potent γ-secretase modulators. Bioorg Med Chem Lett 20:1269–1271
105. Rivkin A, Ahearn SP, Chichetti SM, Hamblett CL, Garcia Y, Martinez M, Hubbs JL, Reutershan MH, Daniels MH, Siliphaivanh P, Otte KM, Li C, Rosenau A, Surdi LM, Jung J, Hughes BL, Crispino JL, Nikov GN, Middleton RE, Moxham CM, Szewczak AA, Shah S, Moy LY, Kenific CM, Tanga F, Cruz JC, Andrade P, Angagaw MH, Shomer NH, Miller T, Munoz B, Shearman MS (2010) Purine derivatives as potent γ-secretase modulators. Bioorg Med Chem Lett 20:2279–2282
106. Bischoff FP, Velter AI, Van BSFA, Berthelot DJ-C (2012) Preparation of triazolylpiperazine derivatives and analogs for use as gamma secretase modulators. WO2012126984A1
107. Wan Z, Hall A, Jin Y, Xiang J-N, Yang E, Eatherton A, Smith B, Yang G, Yu H, Wang J, Ye L, Lau L-F, Yang T, Mitchell W, Cai W, Zhang X, Sang Y, Wang Y, Tong Z, Cheng Z, Hussain I, Elliott JD, Matsuoka Y (2011) Pyridazine-derived γ-secretase modulators. Bioorg Med Chem Lett 21:4016–4019
108. Wan Z, Hall A, Sang Y, Xiang J-N, Yang E, Smith B, Harrison DC, Yang G, Yu H, Price HS, Wang J, Hawkins J, Lau L-F, Johnson MR, Li T, Zhao W, Mitchell WL, Su X, Zhang X, Zhou Y, Jin Y, Tong Z, Cheng Z, Hussain I, Elliott JD, Matsuoka Y (2011) Pyridine-derived γ-secretase modulators. Bioorg Med Chem Lett 21:4832–4835
109. Huang Y, Li T, Eatherton A, Mitchell WL, Rong N, Ye L, Yang X-J, Jin S, Ding Y, Zhang J, Li Y, Wu Y, Jin Y, Sang Y, Cheng Z, Browne ER, Harrison DC, Hussain I, Wan Z, Hall A, Lau L-F, Matsuoka Y (2013) Orally bioavailable and brain-penetrant pyridazine and pyridine-derived γ-secretase modulators reduced amyloidogenic Aβ peptides in vivo. Neuropharmacology 70:278–286
110. Hughes JD, Blagg J, Price DA, Bailey S, Decrescenzo GA, Devraj RV, Ellsworth E, Fobian YM, Gibbs ME, Gilles RW, Greene N, Huang E, Krieger-Burke T, Loesel J, Wager T, Whiteley L, Zhang Y (2008) Physiochemical drug properties associated with in vivo toxicological outcomes. Bioorg Med Chem Lett 18:4872–4875

111. Gijsen HJM, Mercken M (2012) γ-Secretase modulators: can we combine potency with safety? Int J Alzheimers Dis 2012:295207
112. Oehlrich D, Rombouts FJR, Berthelot D, Bischoff FP, De CMAJ, Jaroskova L, Macdonald G, Mercken M, Surkyn M, Trabanco AA, Tresadern G, Van BS, Velter AI, Wu T, Gijsen HJM (2013) Design and synthesis of bicyclic heterocycles as potent γ-secretase modulators. Bioorg Med Chem Lett 23:4794–4800
113. Velter AI, Bischoff FP, Berthelot D, De Cleyn M, Oehlrich D, Jaroskova L, Macdonald G, Minne G, Pieters S, Rombouts F, Van Brandt S, Van Roosbroeck Y, Surkyn M, Trabanco AA, Tresadern G, Wu T, Borghys H, Mercken M, Masungi C, Gijsen H (2014) Anilinotriazoles as potent gamma secretase modulators. Bioorg Med Chem Lett 24:5805–5813
114. Minne GB, Bischoff FP, Gijsen HJM, Velter AI, Pieters SMA, Berthelot DJ-C (2013) Preparation of substituted indole derivatives and related compounds as gamma secretase modulators. WO2013010904A1
115. Takai T, Hoashi Y, Tomata Y, Morimoto S, Nakamura M, Watanabe T, Igari T, Koike T (2015) Discovery of novel 5,6,7,8-tetrahydro[1,2,4]triazolo[4,3-a]pyridine derivatives as gamma-secretase modulators. Bioorg Med Chem Lett 25:4245–4249
116. Fischer C, Zultanski SL, Zhou H, Methot JL, Shah S, Hayashi I, Hughes BL, Moxham CM, Bays NW, Smotrov N, Hill AD, Pan BS, Wu Z, Moy LY, Tanga F, Kenific C, Cruz JC, Walker D, Bouthillette M, Nikov GN, Deshmukh SV, Jeliazkova-Mecheva VV, Diaz D, Michener MS, Cook JJ, Munoz B, Shearman MS (2015) Discovery of novel triazolobenzazepinones as gamma-secretase modulators with central Abeta42 lowering in rodents and rhesus monkeys. Bioorg Med Chem Lett 25:3488–3494
117. Methot JL, Fischer C, Li C, Rivkin A, Ahearn SP, Brown WC, Kattar S, Kelley E, Mampreian DM, Schell A, Rosenau A, Zhou H, Ball R, Deshmukh SV, Jeliazkova-Mecheva VV, Diaz D, Moy LY, Kenific CM, Moxham C, Shah S, Nuthall H, Szewczak AA, Hill A, Hughes B, Smotrov N, Munoz B, Miller TA, Shearman MS (2015) Potent benzoazepinone gamma-secretase modulators with improved bioavailability. Bioorg Med Chem Lett 25:3495–3500
118. Kobayashi T, Iwama S, Fusano A, Kato Y, Ikeda A, Teranishi Y, Nishihara A, Tobe M (2014) Design and synthesis of an aminopiperidine series of gamma-secretase modulators. Bioorg Med Chem Lett 24:378–381
119. Chen JJ, Qian W, Biswas K, Yuan C, Amegadzie A, Liu Q, Nixey T, Zhu J, Ncube M, Rzasa RM, Chavez F Jr, Chen N, DeMorin F, Rumfelt S, Tegley CM, Allen JR, Hitchcock S, Hungate R, Bartberger MD, Zalameda L, Liu Y, McCarter JD, Zhang J, Zhu L, Babu-Khan S, Luo Y, Bradley J, Wen PH, Reid DL, Koegler F, Dean C Jr, Hickman D, Correll TL, Williamson T, Wood S (2013) Discovery of 2-methylpyridine-based biaryl amides as γ-secretase modulators for the treatment of Alzheimer's disease. Bioorg Med Chem Lett 23:6447–6454
120. Takai T, Koike T, Honda E, Kajita Y, Nakamura M, Morimoto S, Hoashi Y, Kamata M, Watanabe T, Igari T, Terauchi J (2015) Design and synthesis of piperazine derivatives as a novel class of gamma-secretase modulators that selectively lower Abeta(4)(2) production. Bioorg Med Chem 23:1923–1934
121. Tran TP, Mullins PB, am Ende CW, Pettersson M (2013) Synthesis of pyridopyrazine-1,6-diones from 6-hydroxypicolinic acids via a one-pot coupling/cyclization reaction. Org Lett 15:642–645
122. Pettersson M, Johnson DS, Subramanyam C, Bales KR, am Ende CW, Fish BA, Green ME, Kauffman GW, Mullins PB, Navaratnam T, Sakya SM, Stiff CM, Tran TP, Xie L, Zhang L, Pustilnik LR, Vetelino BC, Wood KM, Pozdnyakov N, Verhoest PR, O'Donnell CJ (2014) Design, synthesis, and pharmacological evaluation of a novel series of pyridopyrazine-1,6-dione gamma-secretase modulators. J Med Chem 57:1046–1062
123. Pettersson M, Johnson DS, Humphrey JM, Butler TW, am Ende CW, Fish BA, Green ME, Kauffman GW, Mullins PB, O'Donnell CJ, Stepan AF, Stiff CM, Subramanyam C, Tran TP, Vetelino BC, Yang E, Xie L, Bales KR, Pustilnik LR, Steyn SJ, Wood KM, Verhoest PR (2015) Design of pyridopyrazine-1,6-dione gamma-secretase modulators that align potency, MDR efflux ratio, and metabolic stability. ACS Med Chem Lett 6:596–601

124. Pettersson M, Johnson DS, Humphrey JM, am Ende CW, Evrard E, Efremov I, Kauffman GW, Stepan AF, Stiff CM, Xie L, Bales KR, Hajos-Korcsok E, Murrey HE, Pustilnik LR, Steyn SJ, Wood KM, Verhoest PR (2015) Discovery of indole-derived pyridopyrazine-1,6-dione gamma-secretase modulators that target presenilin. Bioorg Med Chem Lett 25:908–913
125. Pettersson M, Johnson DS, Rankic DA, Kauffman GW, Am Ende CW, Butler TW, Boscoe B, Evrard E, Helal CJ, Humphrey JM, Stepan AF, Stiff CM, Yang E, Xie L, Bales KR, Hajos-Korcsok E, Jenkinson S, Pettersen B, Pustilnik LR, Ramirez DS, Steyn SJ, Wood KM, Verhoest PR (2016) Discovery of cyclopropyl chromane-derived pyridopyrazine-1,6-dione γ-secretase modulators with robust central efficacy. MedChemComm. doi:10.1039/C6MD00406G
126. Stepan AF, Kauffman GW, Keefer CE, Verhoest PR, Edwards M (2013) Evaluating the differences in cycloalkyl ether metabolism using the design parameter "lipophilic metabolism efficiency" (LipMetE) and a matched molecular pairs analysis. J Med Chem 56:6985–6990
127. Am Ende CW, Green ME, Johnson DS, Kauffman GW, O'Donnell CJ, Patel NC, Pettersson MY, Stepan AF, Stiff CM, Subramanyam C, Tran TP, Verhoest PR (2014) Preparation of novel bicyclic pyridinones as γ-secretase modulators. WO2014045156A1
128. Pettersson MY, am Ende CW, Johnson DS, Kauffman GW, Stepan AF, Verhoest PR (2016) Preparation of novel cyclopropabenzofuranyl pyridopyrazinediones as γ-secretase modulators for treatment of neurodegenerative diseases. US20160222007
129. Rombouts FJR, Trabanco-Suarez AA, Gijsen HJM, Macdonald GJ, Bischoff FP, Alonso-De Diego S-A, Velter AI, Van Roosbroeck YEM (2013) Preparation of substituted 3,4-dihydro-2H-pyrido[1,2-a]pyrazine-1,6-dione derivatives useful for the treatment of Alzheimer's disease. WO2013171712A1
130. Bischoff FP, Velter AI, Rombouts FJR, De Cleyn MAJ, Van Brandt SFA, Gijsen HJM, Zavattaro C, Van den Keybus FAM (2014) Preparation of novel substituted pyridopiperazinone derivatives as gamma secretase modulators. WO2014111457A1
131. Bischoff FP, Gijsen HJM, Van den Keybus FAM, Rombouts FJR (2014) Novel tricyclic 3,4-dihydro-2H-pyrido[1,2-a]pyrazine-1,6-dione derivatives as gamma secretase modulators and their preparation. WO2014096212A1
132. Qiu RLR, Wills A-M, He P, Leurent C, Hajos-Korcsok E, Mendes da Costa L, Alexander R (2016) PF-06648671 a novel gamma secretase modulator: safety, tolerability, pharmacokinetics and effects on plasma amyloid-β levels following single oral ascending doses in healthy volunteers. In: AAIC Conference, Toronto, Canada, July 22-28, 2016
133. Findeis MA, Schroeder F, McKee TD, Yager D, Fraering PC, Creaser SP, Austin WF, Clardy J, Wang R, Selkoe D, Eckman CB (2012) Discovery of a novel pharmacological and structural class of gamma secretase modulators derived from the extract of Actaea racemosa. ACS Chem Neurosci 3:941–951
134. Fuller NO, Hubbs JL, Austin WF, Creaser SP, McKee TD, Loureiro RM, Tate B, Xia W, Ives JL, Findeis MA, Bronk BS (2012) Initial optimization of a new series of gamma-secretase modulators derived from a triterpene glycoside. ACS Med Chem Lett 3:908–913
135. Hubbs JL, Fuller NO, Austin WF, Shen R, Creaser SP, McKee TD, Loureiro RM, Tate B, Xia W, Ives J, Bronk BS (2012) Optimization of a natural product-based class of gamma-secretase modulators. J Med Chem 55:9270–9282
136. Loureiro RM, Dumin JA, McKee TD, Austin WF, Fuller NO, Hubbs JL, Shen R, Jonker J, Ives J, Bronk BS, Tate B (2013) Efficacy of SPI-1865, a novel gamma-secretase modulator, in multiple rodent models. Alzheimers Res Ther 5:19
137. Hubbs JL, Fuller NO, Austin WF, Shen R, Ma J, Gong Z, Li J, McKee TD, Loureiro RM, Tate B, Dumin JA, Ives J, Bronk BS (2015) Minimization of drug-drug interaction risk and candidate selection in a natural product-based class of gamma-secretase modulators. Bioorg Med Chem Lett 25:1621–1626
138. Jung JI, Ladd TB, Kukar T, Price AR, Moore BD, Koo EH, Golde TE, Felsenstein KM (2013) Steroids as gamma-secretase modulators. FASEB J 27:3775–3785

139. Haass C, Selkoe DJ (2007) Soluble protein oligomers in neurodegeneration: lessons from the Alzheimer's amyloid beta-peptide. Nat Rev Mol Cell Biol 8:101–112
140. Benilova I, Karran E, De Strooper B (2012) The toxic Abeta oligomer and Alzheimer's disease: an emperor in need of clothes. Nat Neurosci 15:349–357
141. Sevigny J, Chiao P, Bussiere T, Weinreb PH, Williams L, Maier M, Dunstan R, Salloway S, Chen T, Ling Y, O'Gorman J, Qian F, Arastu M, Li M, Chollate S, Brennan MS, Quintero-Monzon O, Scannevin RH, Arnold HM, Engber T, Rhodes K, Ferrero J, Hang Y, Mikulskis A, Grimm J, Hock C, Nitsch RM, Sandrock A (2016) The antibody aducanumab reduces Abeta plaques in Alzheimer's disease. Nature 537:50–56

Top Med Chem (2017) 24: 119–158
DOI: 10.1007/7355_2016_17

Published online: 1 March 2017

Inhibitors of Tau-Phosphorylating Kinases

Anna Lucia Fallacara, Iuni Margaret Laura Trist, Silvia Schenone, and Maurizio Botta

Abstract The phosphorylation of tau protein is finely regulated by a balance between phosphorylation and dephosphorylation processes carried out by kinases and phosphatases. It has been suggested that the disruption of this equilibrium and consequent abnormal tau phosphorylation contribute to the aggregation of tau. The understanding of this important mechanism is of high interest because of the implication of tau aggregates in the development of Alzheimer's disease (AD). In the last few years, among the possible strategies which could be used to reduce tau phosphorylation, the inhibition of certain tyrosine kinases has been suggested as a promising alternative to the common therapeutic approaches. In this chapter we will first give an overview of the tau protein kinases, their roles in cells, regulation and importance in AD. This will be followed by a more detailed description of the role of Fyn, a member of the Src family kinases, in the physiological development of CNS and the pathological progress of AD. How the inhibition of Fyn could be used as a new strategy in the fight against AD will be discussed.

Keywords Alzheimer's disease, CNS, Fyn, Inhibitors, Tyrosine kinases

---

A.L. Fallacara and I.M.L. Trist
Dipartimento di Biotecnologie, Chimica e Farmacia, Università degli Studi di Siena, Via Aldo Moro 2, 53100 Siena, Italy

S. Schenone
Dipartimento di Farmacia, Università degli Studi di Genova, Viale Benedetto XV, 3, 16132 Genoa, Italy

M. Botta (✉)
Dipartimento di Biotecnologie, Chimica e Farmacia, Università degli Studi di Siena, Via Aldo Moro 2, 53100 Siena, Italy

Sbarro Institute for Cancer Research and Molecular Medicine, Center for Biotechnology, College of Science and Technology, Temple University, BioLife Science Building, Suite 333, 1900 N 12th Street, Philadelphia, PA 19122, USA
e-mail: botta.maurizio@gmail.com

Contents

Abbreviations

5-HT6R	5-Hydroxytryptamine 6 receptor
AD	Alzheimer's disease
ADRA	Alzheimer's Disease and Related Disorders Association
AMPAR	α-Amino-3-hydroxy-5-methyl-4-isoxazolepropionic acid receptor
APP	Amyloid precursor protein
Aβ	Amyloid β
Blk	B lymphocyte kinase
CaMKII	Ca^{2+}/calmodulin-dependent protein kinase II
CDK5	Cyclin-dependent protein kinase-5
CHK1	Checkpoint kinase 1
CIs	Cholinesterase inhibitors
CK	Casein kinase
c-KitR	c-Kit receptor
CNS	Central nervous system
CSF	Cerebrospinal fluid
Csk	C-terminal Src kinase
DYRK	Dual-specificity tyrosine phosphorylation-regulated kinase

ECM	Extracellular matrix
EGFR	Epidermal growth factor receptor
Erk	Signal-regulated kinase
FAD	Familial Alzheimer's disease
Fgr	Gardner-Rasheed feline sarcoma viral (v-fgr) oncogene homolog
Frk	Fyn-related kinase
GABA(A)R	γ-Aminobutyric acid A receptor
GAP	GTPase-activating protein
GSK3	Glycogen synthase kinase-3
HRI	Hydrophobic region I
HRII	Hydrophobic region II
IGF-1R	Insulin-like growth factor 1 receptor
JAK2	Janus kinase 2
JNK	c-Jun N-terminal kinase
KDR	Kinase insert domain receptor
Lck	Lymphocyte-specific tyrosine kinase
MAP	Microtubule-associated protein
MAPK	Mitogen-activated protein kinases
MARK	Microtubule affinity-regulating kinases
MBD	Microtubule-binding domain
MCI	Mild cognitive impairment
MCT	Mast cell tumours
MD	Molecular dynamics
MMGBSA	Molecular mechanics Generalised born surface area
mTOR	Mammalian Target of Rapamycin
NFT	Neurofibrillary tangles
NIA	National Institute of Ageing
NINCDS	National Institute of Neurological and Communicative Disorders and Stroke
NMDA	*N*-methyl-D-aspartate
NMDAR	NMDA receptor
Non-PDPK	Non-proline-directed protein kinases
nRTK	Non-receptor tyrosine kinase
NSAIDs	Nonsteroidal anti-inflammatory drugs
PDGFR	Platelet-derived growth factor receptor
PDPK	Proline-directed protein kinases
PhK	Phosphorylase kinase
PKA	Protein kinase A
PKB	Protein kinase B
PKC	Protein kinase C
PKN	Protein kinase N
PrP	Cellular prion protein
PSD	Postsynaptic density
PyrPyr	Pyrazolo[3,4-d]pyrimidine

RPTP	Receptor protein tyrosine phosphatase
SFK	Src family kinase
SP	Senile plaques
Syk	Spleen tyrosine kinase
TK	Tyrosine kinase
TPK	Tyrosine protein kinases
TTBK	Tau-tubulin kinase
WHO	World Health Organization

1 Alzheimer's Disease (AD)

As defined by the National Institute of Ageing (NIA), "Alzheimer's disease (AD) is an irreversible, progressive brain disorder that slowly destroys memory and thinking skills and, eventually, the ability to carry out the simplest tasks" [1]. AD was named in 1910 after the German neurologist Alois Alzheimer, who reported for the first time the case of a patient with symptoms that were unusual at the time: reduced memory and comprehension, disorientation, aphasia, paranoia, unpredictable behaviour, auditory hallucination and pronounced psychosocial impairment [2]. After the patient's death, Dr. Alzheimer could observe brain atrophy and the presence of "dense bundles of fibrils" as well as of "minute miliary foci which are caused by the deposition of a special substance" distributed all over the cortex [3]. Indeed, it is now well known that AD is a neurodegenerative disease characterised by the presence of these two types of hallmarks. The "dense bundles of fibrils" are neurofibrillary tangles (NFT): intraneuronal aggregations, the main constituent of which is abnormally phosphorylated tau protein [4]. The "miliary foci", also called senile plaques (SP), are extracellular deposits of amyloid β (Aβ) proteins. In addition to these, loss of connections between neurons in the brain, inflammation, severe tissue shrinkage and eventual death of brain cells are often considered features of AD [1, 5].

1.1 Epidemiology

In developed countries, birth control has led to a general decline in fertility, accompanied by an enhancement of the lifestyle that improves longevity. Developing countries are following the same trend. Fewer children are entering the world population, and people are living longer, making the proportion of old people increase from 7 to 12% in 2015 [6, 7]. According to the World Health Organization (WHO), the pace of population ageing is much faster than in the past, and it is estimated that in 2050 the proportion of the world's population over 60 years will be approximately 22% [6]. Given that AD generally becomes apparent after the mid-60 years of age (see Fig. 1) [1, 8], it is not surprising that AD prevalence is

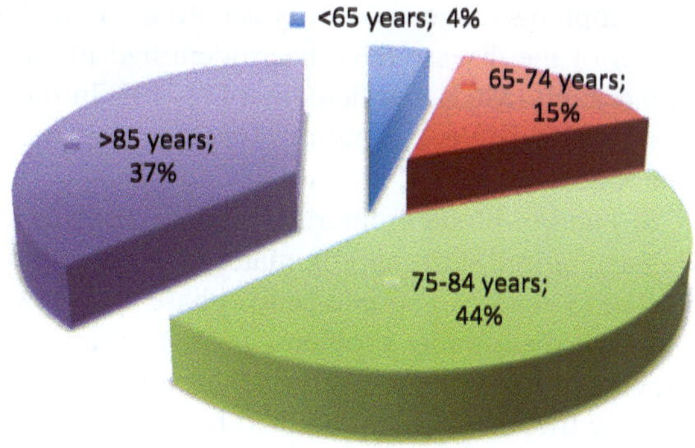

Fig. 1 Age distribution of people with AD in the USA in 2016. Data from [9]

significantly increasing, with a dramatic impact on public health and healthcare financing, mainly because of its requirement of persistent caregiving [1, 6, 7]. Indeed, in the USA alone, over five million people are currently living with AD, and experts say that this will cost over $236 billion in 2016, making AD one of the potential most expensive chronic diseases [5, 8]. It is clear that the development of a cure for AD is compelling.

1.2 Symptoms and Disease Progression

Criteria for clinical diagnosis for AD were first established only in 1984 by the National Institute of Neurological and Communicative Disorders and Stroke (NINCDS) and the Alzheimer's Disease and Related Disorders Association (ADRA) [10]. Due to brain plasticity, the symptoms of AD usually appear approximately a decade after the beginning of brain damage. Symptoms usually affect people in different ways, and the criteria for AD diagnosis are regularly updated [5, 8]. Since brain damage usually involves regions responsible of forming new memories, the first sign of AD is the gradual worsening of the ability to remember new information [1, 8]. With disease progression, other signs that are indicative of AD include [11–13]:

- Memory loss disrupting daily life
- Challenges in planning or solving problems
- Confusion with time or place
- Difficulty in completing familiar tasks
- Increasing problems with words in speaking or writing
- Trouble in understanding visual images and spatial relationships
- Decreased or poor judgement
- Loss of the ability to retrace steps
- Misplacing things
- Changes in mood and personality
- Withdrawal from work or societal activities

Symptoms range in severity according to the level of damage of the brain. Three stages of the disease have been identified to date: preclinical AD, mild cognitive impairment (MCI) and dementia [11–13]. In preclinical AD, patients have not yet developed visible symptoms, for example, memory loss, and the disease goes unnoticed. However, measurable changes in the brain and cerebrospinal fluid (CSF) and/or blood biomarkers can be detected [14]. As already mentioned, brain damage indeed starts approximately a decade before the appearance of signs [1, 14].

The symptomatic predementia phase of AD (MCI due to AD) is characterised by mild but measurable problems in performing complex functional tasks that were previously executed regularly. However, this does not impact on the independence in daily life, and only minimal aid or assistance is needed [13]. In the last phase of AD, dementia, there is a more important impairment in the patient's memory, thinking and behaviour, decreasing the ability to be independent in everyday life. In this case, continuous care is needed [12].

1.3 Causes of AD

The exact mechanism behind the brain lesions that cause AD is still not completely understood. It is plausible that for most patients there is a combination of genetic, environmental, lifestyle factors that sum to the age-related changes in the brain (e.g. atrophy, inflammation, production of free radicals or mitochondrial dysfunction) [1, 5].

Nevertheless, in 1–5% of AD cases, there is an early onset of the symptoms (before 60 years of age, Fig. 1), and the disease itself progresses at a faster rate. This condition, known as familial AD (FAD), is inherited in an autosomal dominant fashion of mutated genes (amyloid precursor protein, presenilin 1 and presenilin 2) that lead to Aβ aggregation [15, 16].

1.4 AD Aetiology

In AD, the dysfunction and cognitive impairment in patients are closely associated with synaptic loss, neurotransmitter depletion and brain lesions, in particular in regions associated with memory and cognition [17–20]. These lesions result in a multiplicity of irregular protein changes that occur in the brain. To date, the complete picture of the molecular and cellular mechanisms that are behind these cascades is still not clear and sometimes controversial. However, the increasing understanding of the complex brain changes involved in the onset and progression of AD allows the development of several and different intervention strategies focused on targeting the causes of neuronal cell death [1, 20, 21]. The main hypotheses on the causes of AD are the cholinergic hypothesis, the glutamatergic hypothesis, the amyloid hypothesis and the tau hypothesis. An important aspect of the disease is the consistent finding of abnormal inflammatory responses in AD brains, with upregulation of pro-inflammatory cytokines in the CSF, as confirmed by the fact that the regular use of nonsteroidal anti-inflammatory drugs (NSAIDs) has been correlated to a decreased risk of developing AD [22–24].

1.4.1 Cholinergic Hypothesis

This theory was first proposed in 1982 and is based on the relation between dementia symptoms and impaired neurotransmission and degeneration of neuronal circuits, in particular of the cholinergic system [25]. The functional disturbances in cholinergic activity, for example, the progressive loss of cholinergic neurons or the reduced activity of choline acetyltransferase in neocortex and hippocampus, produce a decline in levels of acetylcholine in the brain and play a role in memory loss and related cognitive problems [26, 27].

1.4.2 Glutamatergic Hypothesis

N-Methyl-D-aspartate (NMDA) receptors are ionotropic glutamate channels that are inhibited by Mg^{2+} at resting potential and activated by glutamate, the main excitatory neurotransmitter in the central nervous system (CNS). AD patients present high levels of glutamate that over-activate the NMDA channels, leading to excessive Ca^{2+} influx and cell death [28].

1.4.3 Amyloid Hypothesis

Probably one of the most studied hypotheses for AD, the amyloid hypothesis, was first presented in 1991 [29, 30]. In this case, sequential proteolysis of the amyloid precursor protein (APP) produces the Aβ peptides, which accumulate into toxic protein aggregates that cause the formation of the SP, with consequent synaptic failure and neurodegeneration [31]. APP is cleaved following two possible pathways (the non-amyloidogenic and the amyloidogenic), and predisposition for AD has been observed when the second one is prevalent over the first one. In the amyloidogenic pathway, the cleavage of APP is heterogeneous and the produced Aβ peptides are different in length. Among these, the hydrophobic $A\beta_{42}$ is more aggregation prone and is considered more toxic [31–34].

1.4.4 Tau Hypothesis

One limitation of the amyloid hypothesis is that it does not explain the presence of the NFT in AD brains but only of SP. On the other hand, NFT can be explained by the tau hypothesis, and indeed AD is often accounted as a "tauopathy" [35]. Tau is abundantly expressed in the brain and belongs to the microtubule-associated protein (MAP) family. Its ability to bind to microtubules makes tau crucial in the stabilisation of neuronal cytoskeleton [35–37]. As further explained in Sect. 2, the hyperphosphorylation of this protein induces a decrease in tau-microtubule interaction, trigging a destabilisation of the neuronal cytoskeleton and tau accumulation [36–40].

1.5 Current Treatment of AD

Despite the increasing prevalence of AD, there is still no cure, but only symptomatic therapy [1, 5, 41, 42]. The available drugs (donepezil [43], galantamine [44], memantine [45], rivastigmine [46]) and a recently approved drug combination (donepezil with memantine [47]) aim to help patients to maintain mental function, to manage behavioural symptoms and, in general, to slow or delay the progression of the disease [41, 42]. These compounds are classified in two groups: cholinesterase inhibitors (CIs) and NMDA antagonist. Donepezil (Pfizer), galantamine (Janssen) and rivastigmine (Novartis) belong to the first group and are regarded as standard and first-line treatment for all stages of AD [41]. As explained in Sect. 1.4.1, the loss of cholinergic transmission has been correlated with AD symptoms. Thus, the inhibition of the degradation of acetylcholine improves the functional disturbances in cholinergic activity, with a consequent delay in the decline in cognitive function [43, 48]. Memantine (Lundbeck) belongs to the NMDA antagonists family. It is usually prescribed for moderate-to-severe AD and is believed to protect neurons from the excitotoxicity of NMDA channel over-activation (see Sect. 1.4.2). This compound improves cognition, and it may reduce behavioural and psychological symptoms [49]. The combination of a CI (donepezil) and a NMDA antagonist (memantine) produces a significant improvement in cognitive function, language, behaviours and global state in moderate-to-severe AD, but less in mild-to-moderate AD [5, 47]. Besides the FDA-approved therapies, current suggestions for prevention of AD include the use of antioxidants, omega fatty acid supplements, physical activity and cognitive engagement [42].

2 Implications of Tau Phosphorylation in AD

Tau is a 55-kDa protein expressed mainly in neurons and presents four regions: an acidic portion at the N-terminal, a proline-rich region, a microtubule-binding domain (MBD) and a C-terminal. The MBD contains repeat domains with a conserved KXGS motif and is consequently also called the repeat region [32, 36–38]. In the adult brain, tau exists in six isoforms that are formed through alternative mRNA splicing of the same gene on chromosome 17q. The six isoforms mainly differ in the absence/presence of one of the acidic inserts in the N-terminal acidic region and in the number (three or four) of repeat domains in the MBD [36, 50]. Since the repeat domains are responsible for the binding of tau to microtubules, isoforms containing four repeat domains bind with greater affinity to microtubules than do the ones containing only three repeat domains [36].

Up to 85 phosphorylation sites have been identified in tau. Among them, 28 are phosphorylated only in AD brains, 16 are phosphorylated both in control and AD brains and 31 are phosphorylated at physiological conditions (Table 1) [37]. Phosphorylation can include the serine (Ser) residues of the KXGS motif or threonine (Thr) residues in the repeat domains (residues in red in Table 1). Within these amino acids, the phosphorylation of specific sites (262, 293, 324, 356, highlighted in Table 1) produces a change in tau conformation with consequent decrease in tau binding to microtubules [36–38]. Thus, on one hand, hyperphosphorylation of tau produces a general cytoskeleton destabilisation with a disruption of neuronal functions that involve axonal growth, vesicle and organelle transport and nervous signal propagation. On the other hand, hyperphosphorylated tau proteins are less susceptible to protease processing and accumulate in cells as paired helical filaments (PHFs), the unit fibrils of NFT, and eventually produce cell death [36–39, 51]. The association of NFT with AD symptoms was confirmed by the positive correlation between the number of NFTs and the severity of dementia [52].

3 Kinases Involved in Tau Phosphorylation

The kinases responsible for tau phosphorylation are clustered in three groups (Fig. 2): proline-directed protein kinases (PDPK), non-proline-directed protein kinases (non-PDPK) and tyrosine protein kinases (TPK) [37].

3.1 Proline-Directed Protein Kinases, PDPKs

PDPKs phosphorylate a serine or threonine residue that precedes a proline amino acid. Three main groups of kinases phosphorylate tau by recognising this S/TP

Table 1 Tau phosphorylation sites

Tau phosphorylation sites found in:					
Only AD brain		**AD and control brains**	**Physiologic conditions**		**Putative sites**
Y18	**S258**	S46	T17	S241	T30
S68	**S262**	T181	**Y29**	**T245**	S61
T71	S289	S198	T39	T263	T63
S113	**S356**	S199	T50	**S285**	S64
T123	**Y394**	S202	T52	**S293**	T76
T153	T403	T205	S56	**S305**	S129
T175	S409	T212	T95	S316	S137
T184	S422	T217	T101	**S320**	**Y310**
S185	T427	T231	T102	**S324**	T319
S191	S433	S235	T111	S431	T377
Y197	S435	S396	S131	**S352**	
S208		S400	T135	**T361**	
S210		S404	T149	T373	
S214		S412	T169	T386	
S237		S413	S195	T386	
S238		S416	T220		

Adapted from [36, 37]. See text for details

motif: glycogen synthase kinase-3 (GSK3), cyclin-dependent protein kinase-5 (CDK5) and mitogen-activated protein kinases (MAPK) [37].

3.1.1 GSK3

GSK3 was initially discovered for its involvement in glycogen metabolism and is one of the most exploited kinase targets in AD. It plays roles in general neuronal functions, apoptosis, cell proliferation, oncogenesis, embryonic development and immune response [53–57]. There are two main GSK3 isoforms (GSK3α and GSK3β) sharing 85% sequence homology, but the isoform associated to tau

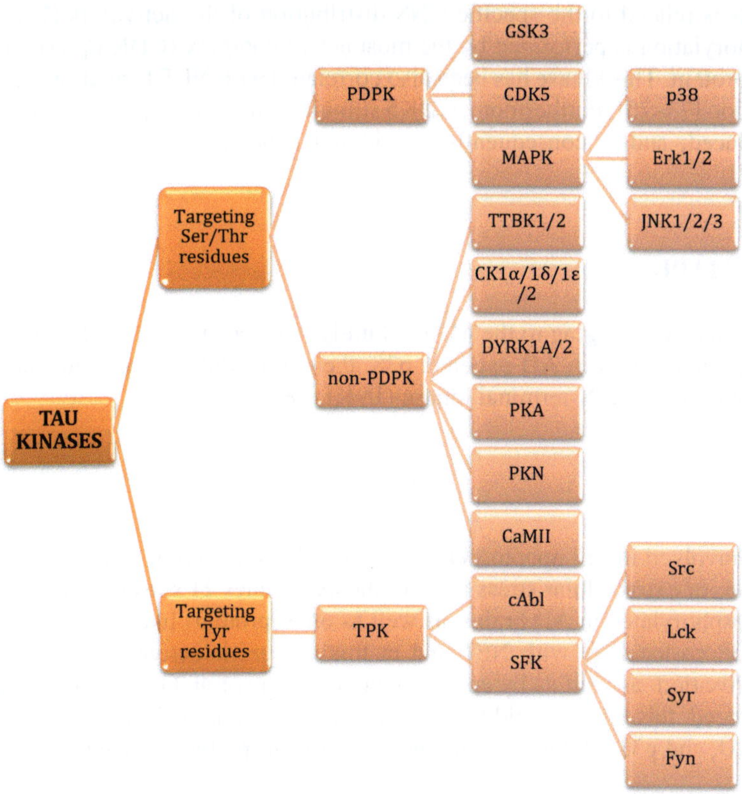

Fig. 2 Summary of tau-phosphorylating kinases

hyperphosphorylation is GSK3β [58, 59]. GSK3β can phosphorylate tau at 42 sites, of which 29 are phosphorylated in AD brains [60]. Overexpression or increased activity of GSK3β in mouse models induces tau hyperphosphorylation and other tau-mediated effects, for example, neurodegeneration, inflammatory response, increased production of Aβ and reduced acetylcholine synthesis [61–63]. GSK3β co-localises with NFT, and its increased activity is observed in AD brains, making this a promising therapeutic target [55, 64–66].

3.1.2 CDK5

CDK5 is pivotal in the development of the CNS as it is involved in several key processes, like the regulation of synaptic functions, neuronal cytoskeleton dynamics and neuritic development [67–70]. The monomeric form of CDK5 is enzymatically inactive and needs to interact with a co-activator (namely, p39/p35 and p29/p25) to phosphorylate substrates. Thus, the restriction of CDK5 activity to neurons is related to the specific CNS distribution of the activators [71, 72]. Tau phosphorylation is performed by the most active complex (CDK5/p25) on 11 sites of the protein. This kinase has been also correlated with NFT formation and GSK3β activation [73–75]. Furthermore, CDK5 stimulates neuronal apoptosis, in particular Aβ-induced, and is consequently considered a potential target for AD therapy [76, 77].

3.1.3 MAPK

Three kinases belonging to the MAPK family have been associated with abnormal tau phosphorylation in AD patients: p38, the extracellular signal-regulated kinase (Erk) and the c-Jun N-terminal kinase (JNK) [78, 79].

P38

P38 is involved in cell apoptosis and is generally activated in response to cytokine exposure or extracellular stress [80]. It phosphorylates 21 sites of tau, 15 of which are phosphorylated in AD brains, as well as other proteins like p53 and GSK3β. In AD, p38 co-localises with NFT and in this context is activated by cytoskeleton destabilisation, as well as by environmental, oxidative and chemical stresses [81–85]. Indeed, this kinase could be involved in the inflammatory response associated with AD, and p38 inhibitors could be used as neuroprotective agents [86].

Erk

Two Erk isoforms (Erk1 and Erk2), sharing 83% sequence homology, have been correlated to AD. Both of them regulate cell growth, proliferation and division, as well as cell death [87, 88]. Erk1/2 can be activated by cytoskeleton abnormalities, and their increased activity produces tau hyperphosphorylation (at 16 possible sites) and apoptosis, becoming a putative drug target [81, 87, 89].

JNK

All three isoforms of JNK (JNK1, JNK2 and JNK3) are expressed in the brain and phosphorylate tau in 12 sites [81]. Similarly to p38, JNK can be activated by cytoskeleton destabilisation and induces apoptosis in response to stress. JNK-activating stresses include anticancer drugs (e.g. etoposide), ultraviolet or ionising radiation, nerve growth factor deprivation or cytokines [90, 91]. As well as co-localising with NFT, JNK is involved in SP formation by inducing γ-secretase activity and consequent Aβ aggregation, making JNK inhibition a possible therapeutic strategy [92].

3.2 Non-PDPK

Several kinases that belong to the non-PDPK family are able to phosphorylate tau: tau-tubulin kinase (TTBK), casein kinase (CK), dual-specificity tyrosine phosphorylation-regulated kinase (DYRK), microtubule affinity-regulating kinases (MARK), phosphorylase kinase (PhK), protein kinase A (PKA), protein kinase B (PKB), protein kinase C (PKC), protein kinase N (PKN) and Ca^{2+}/calmodulin-dependent protein kinase II (CaMKII). Some of them do not represent interesting potential therapeutic targets and will not be further discussed here.

3.2.1 TTBK

Two tau-tubulin kinases (TTBK1 and TTBK2) are encoded by different genes and are expressed in nonoverlapping regions of the brain [93]. In addition to ten possible sites in tau, TTBK1/2 also phosphorylate tubulin. TTBK1, in particular, was found to phosphorylate tau and neurofilaments as well as increase CDK5/p25 complex formation, activate GSK3β and consequently induce NFT formation and neuronal loss [81, 94].

3.2.2 CK

Both CK1 and CK2 are promising targets for the development of new drugs because they phosphorylate tau in 46 sites and, thus, regulate microtubule dynamics, apoptosis and cell survival pathways [95–97]. Of the many different isoforms of these kinases, only some specific ones (CK1α, CK1δ, CK1ε and CK2) are involved in AD and are activated by Aβ. Notably, CK1 isoforms are increased in AD and co-localise with NFT [98–102].

3.2.3 DYRK

Only two of the six DYRK isoforms (DYRK1A and DYRK2) are involved in AD. They both phosphorylate two proteins involved in learning and memory processes: tau and the transcription factor cAMP response element binding [103, 104]. DYRK1A is encoded by a gene on the 21st chromosome and contributes to the positive correlation between Down syndrome and AD, along with the APP gene located in the same chromosome. Moreover, its increased levels in AD brains make this isoform particularly interesting for the development of AD and Down syndrome therapies [105–107].

3.2.4 PKA

PKA is a pro-apoptotic kinase activated by cAMP and responsible for the phosphorylation of several proteins, for example, PKB, PKC, GSK3 and tau [108, 109]. Of the 25 tau phosphorylation sites targeted by PKA, 17 are phosphorylated in AD brains. Indeed, it has been associated with AD tau hyperphosphorylation and co-localises with NFT. PKA is activated in AD, causing memory loss and making it a potential target for AD therapy [110, 111].

3.2.5 PKN

The inhibition of PKN is another possible therapeutic strategy. This protein kinase is highly expressed in the brain and has been co-localised with NFT. It phosphorylates GSK3β as well as tau in four sites, two of which are phosphorylated in AD brains [112–115].

3.2.6 CaMKII

CaMKII kinase is involved in apoptosis induced by retinoic acid and phosphorylates tau in several sites, five of which are phosphorylated in AD brains. When phosphorylated at Thr287, CaMKII binds to the NMDA receptor. This kinase is expressed in neurons with NFT, and the inhibition of this protein could be a promising therapy for AD [116–118].

3.3 Tyrosine Protein Kinases, TPKs

The majority of tau phosphorylated amino acids are serine and threonine residues. However, five sites are represented by tyrosines 18, 29, 197, 310 and 394 (in green in Table 1). Four of these (18, 29, 197 and 394) are phosphorylated only in AD brains, and tyrosine phosphorylation might promote additional phosphorylation at other sites. Consequently, TPK that can phosphorylate tau are of particular therapeutic interest [119–124]. The main non-receptor tyrosine kinases (nRTKs) identified to phosphorylate tau are non-receptor Src family kinases (SFKs) members and cAbl kinase [37].

3.3.1 cAbl

This nRTK regulates cell proliferation and apoptosis through several pathways [123]. It is involved in the first steps of NFT formation, by phosphorylating tau at

Tyr18 and Tyr394 [122]. The activation of cAbl occurs through phosphorylation of Tyr245 that can be mediated by SFKs. In neuronal cells, cAbl activation allows it to bind to p73 and trigger apoptotic signalling. Importantly, in AD both cAbl and p73 are highly expressed and stimulated by Aβ itself [124, 125].

3.3.2 SFK

SFK members involved in tau phosphorylation are mainly Src, lymphocyte-specific tyrosine kinase (Lck) and Fyn [37]. Among these, Fyn has been strongly related to AD symptoms, and its involvement in AD as well as its potential as a novel cure for AD will be further explored later in this manuscript.

4 The Development of Inhibitors of Tau-Phosphorylating Kinases Is a Promising Strategy in AD Drug Development

As already mentioned, there is no available AD disease-modifying therapy, which means that drugs only ameliorate symptoms and do not stop neuronal degeneration. Thus, there is no halt in the progressive deterioration of the cognitive faculties, and only patients presenting mild-to-moderate AD have benefits [41, 126, 127]. The need of novel efficacious therapy has encouraged research on understanding AD aetiology and progression. The increased knowledge has pushed the development of novel disease-modifying strategies, for example, immunotherapy or small molecules involved in amyloid deposition modulation, metal interference, secretase modulation, sirtuins and caspase inhibition, tau deposition modulation, oxidative damage reduction and cholesterol and vascular-related risk factor modulation [37, 41, 126–129]. These novel strategies have already been reviewed elsewhere and will not be further explored, as it is beyond the scope of this manuscript. The success of these approaches would improve the duration and quality of life of AD patients and allow disease intervention at the predementia stage, expanding considerably the market for AD drug development, which was already around $4.1 billion in 2014 [130]. Indeed, today more than 100 clinical trials that have AD as therapeutic indication are ongoing, and a high proportion of them, for example, immunotherapy and secretase modulation, follow the historically most studied amyloid hypothesis (see Sect. 1.4.3) [127, 130]. Unfortunately, despite all the efforts, clinical trials are producing minimal results, and AD drug development in clinical trials has proved to be challenging [131]. On the other hand, the tau hypothesis (see Sect. 1.4.4) explains more aspects of AD-related events. Several encouraging molecules have been developed following this hypothesis, acting at different levels: microtubule stabilisation, tau oligomerisation, tau phosphorylation, tau dephosphorylation, tau degradation and tau vaccination. Most of these agents

Fig. 3 Tideglusib

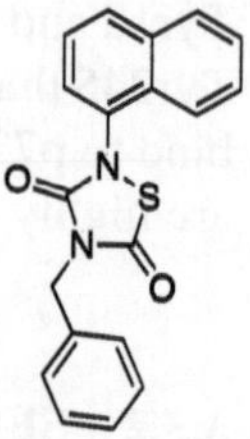

are still in preclinical tests, and the hope is that they will perform better than the current molecules in clinical trials [129].

The strong evidence supporting the contribution of tau-phosphorylating kinases suggests that these enzymes are very promising targets for the development of novel anti-AD agents [126, 132]. The development of protein kinase inhibitors is a long-established strategy in cancer therapy, where inhibitors reduce the hyperactivity of kinases that causes uncontrolled cell division, migration and invasion. These compounds often display limited side effects and are well tolerated in patients, allowing the use of protein kinase inhibitors in AD drug development as well [132–134]. Inhibitors are being developed against almost all the tau-phosphorylating protein kinases mentioned above: GSK3β, CDK5, p38, Erk1/2, JNK, TTBK1/2, CK1/2, DYRK1A, PKA, CaMKII and TPKs [37]. Compounds in preclinical and clinical development have been extensively reviewed elsewhere [37, 126–128, 132]. Among tau-phosphorylating kinases, GSK3β is one of the most studied with respect to AD, and some of its inhibitors have reached clinical development. Of interest, lithium (as Li^+), the classic mood stabiliser used in bipolar depression, was discovered to compete with Mg^{2+} and to inhibit GSK3β selectively. Phase I and II clinical trials are ongoing to investigate the efficacy of this ion in AD (NCT02129348, NCT02204969, NCT02601859) [37, 128]. Furthermore, it has been demonstrated that intranasal insulin has inhibitory effects on GSK3β, and two of these products have reached phase II clinical development: Humulin R for mild AD (NCT02503501) and glulisine for mild-to-moderate AD (NCT01436045) [127]. On the other hand, another very attractive GSK3β inhibitor, tideglusib (Fig. 3), has failed a safety phase I/II clinical trial (NCT00948259) and an efficacy, safety and tolerability phase II study (NCT01350362), and its development was halted [127, 128].

This suggests that there is room for improvement also within the development of tau-kinase inhibitors. In this respect, Fyn represents a very promising drug development target, as it is involved in several aspects of AD pathogenesis, as further analysed later in the manuscript. As a proof of concept, two Fyn inhibitors, masitinib (NCT00976118 and NCT018725998) and saracatinib (NCT01864655 and NCT02167256), have reached clinical development (see Sect. 8) [135].

5 Fyn Kinase: Structure and Functions

Human Fyn is a tyrosine kinase (TK) belonging to the SFKs (Src family kinases), of which c-Src is the prototypical and best-studied family member. This family is constituted by nine members grouped into subclasses: the SrcA subclass includes Src, Yes, Fyn and Fgr and the SrcB subclass includes Lck, Hck, Blk and Lyn, while Frk is in its own subfamily. SFKs have been implicated in most cancer types [136], where they play a critical role in the regulation of cellular processes like proliferation, cell-cell and cell-matrix (ECM) adhesion and angiogenesis [137, 138]. In the CNS Src, Fyn and Yes are widely expressed, while Lck is commonly found in T cells. Blk, Hck, Fgr and Lyn are each expressed primarily in haematopoietic cells, while Fyn-related kinase (Frk) is a nuclear TK, and its functions mostly concern the regulation of G1 and S phases of the cell cycle [139]. Fyn is a 59-kDa protein comprising 537 amino acids encoded by the FYN gene located on chromosome 6q21 [140]. There are three different isoforms of Fyn: isoform 1 (FynB), the first identified; isoform 2 (FynT), which is expressed in T cells and differs from FynB in the linker sequence between SH2 and SH1 domains; and isoform 3, which has been found in blood cells and differs from isoform 1 by deletion of residues 233–287 [141]. Even if the first two forms are expressed as a mixture in different tissues, FynB seems to be highly expressed in the brain, while FynT tends to be expressed in T cells.

Like the other members of SFKs and many other TKs involved in signal transduction, Fyn consists of domains which mediate protein-protein interactions. Typically, five regions have been recognised as important for protein functions (Fig. 4a). These regions are the SH1 domain, the SH2 and SH3 domains, the unique domain and the SH4 domain. SH1 is also known as the catalytic domain and was the first region of homology identified among different Src family members [142]. It shares the characteristic bilobal structure with all tyrosine and ser/thr kinases. The N-terminal (small) lobe of SH1 is composed of five β-strands and one α-helix, termed the C-helix, which is a crucial element in the regulation of the enzyme activity. The C-terminal (large) lobe is constituted by α-helices and contains the activation loop with its DFG (Asp-Phe-Gly) motif. The hinge region connects the two lobes and represents an important anchor point for ATP. The SH2 domain is about 100 amino acids long and is important because it binds the tyrosine-phosphorylated residues in a specific sequence [143, 144].

The SH3 domain is approximately 75 amino acids in length and usually keeps contacts with proline-rich sequences on target substrates. SH3 is involved in the regulation of the autoinhibition mechanism of SFK activity by binding the region which links the SH2 domain to the SH1 domain [145–147]. The unique domain is located at the N-terminus of the kinase and it is subjected to a high variability among the SFKs. The role of the Fyn unique domain has not been completely clarified, but it has been proposed to be required for subcellular localisation [148]. Finally, the N-terminal seven residues of Fyn constitute the SH4 domain, which contains potential sites for post-translational modifications, like fatty

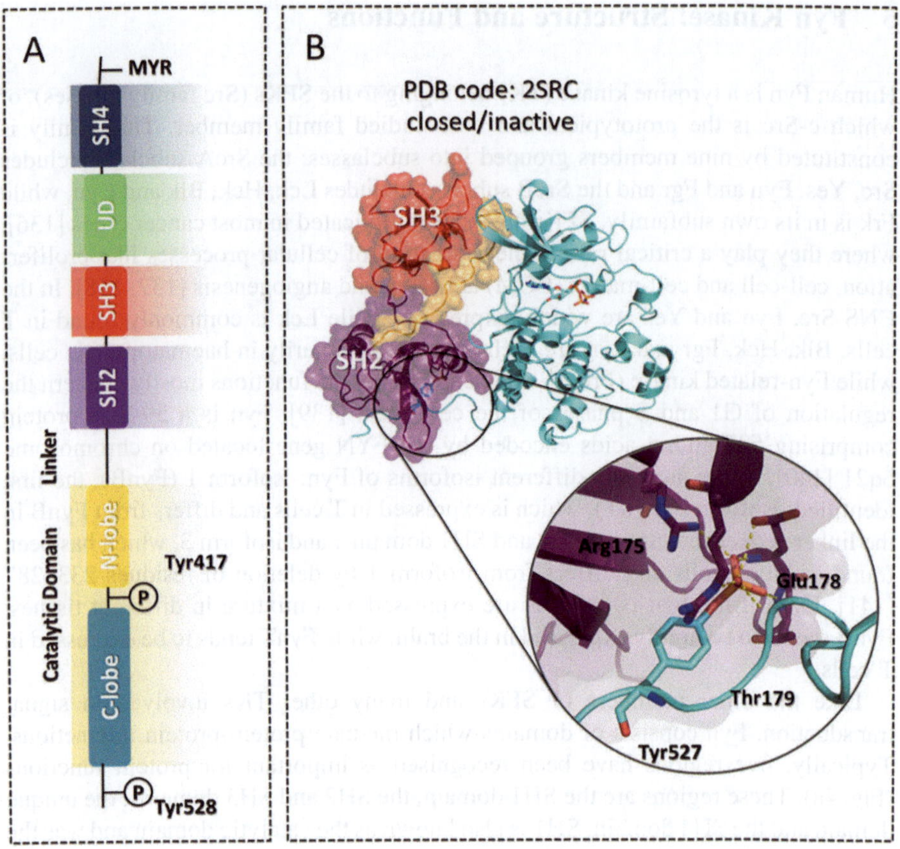

Fig. 4 Panel **A**, schematic representation of Fyn kinase structure. Phosphorylation sites have been highlighted. Panel **B**, X-ray structure of the closed/inactive conformation of c-Src. The binding of Tyr527 to the SH2 domain which stabilises the closed conformation is highlighted in the magnification

acylation, important for association with the plasma membrane [140, 149]. The presence of a glycine residue in position 2 (common in many TK such as Abl) serves, after the removal of the leading methionine, as a site of modification by a myristate group, while cysteine 3, not present in Src and Blk, can be modified with palmitate [150]. These acylations allow the anchorage of the kinase to the inner surface of the cell membrane.

The activation of Fyn depends on the binding of the SH2 and SH3 domains to their ligands and the phosphorylation/dephosphorylation of specific tyrosine residues. In particular, the phosphorylation of Tyr528 on isoform 2 or FynT (residue 531 on FynB) by the C-terminal Src kinase (Csk) inhibits the enzyme, promoting the formation of a closed conformation in which the catalytic site is not accessible to ATP and protein substrates (Fig. 4b). On the other hand, the phosphorylation on

Tyr417 on Fyn T (residue 420 on FynB) promotes Fyn activation and leads to an open/active conformation of the protein structure [151].

6 Functions of Fyn in the Central Nervous System (CNS)

Fyn is considered to be involved both in the development of the CNS and in normal adult brain physiology. The biological importance of Fyn in the brain has been extensively investigated using transgenic animal models. These studies indicated that Fyn is a key element for the development of CNS. Mice deficient in Fyn display various neural defects, including disruption of hippocampal architecture, impaired spatial learning and increased sensitivity to ethanol [152], suggesting the importance of Fyn in various neural signalling pathways. Moreover Fyn plays important roles in CNS myelination, and it has been found in many brain areas [153]. In the last decade, many efforts have been devoted to the study of the role of Fyn in CNS. Fyn has been found in glial cells in white matter and cultured oligodendrocytes [153], where it is important for the formation of neuritis. Indeed, Fyn interacts with γ-tubulin, a member of the tubulin family involved in the nucleation and polar orientation of microtubules. Consistently, the pretreatment of P19 embryonic carcinoma cells with SFK inhibitors blocked the nucleation activity of γ-tubulin [154].

Fyn and other members of SFKs are involved in synaptic transmission and plasticity, and indeed it is localised to the postsynaptic density (PSD), the primary cytoskeleton specialisation at neuronal excitatory synapses, constituted by many proteins such as PSD95, NMDAR and AMPAR. In particular, NMDAR forms a multiprotein complex, with PSD95 directly binding to the NR2 subunit of NMDAR. PSD95 interaction with the SH2 domain of Fyn has been proposed to induce the phosphorylation of NMDAR subunit NR2A [155, 156]. In this way Fyn regulates the NMDA phosphorylation and the NMDAR-dependent synaptic potentiation [157]. It was demonstrated that Fyn phosphorylates PSD93, another protein usually associated with NMDAR. PSD93 serves as an anchor substrate for Fyn, placing Fyn very close to NMDAR, regulating the receptor activity [158].

Fyn has many different roles in the development and regulation of physiological functions of the CNS, and every year new associations with target proteins are discovered.

Jud and colleagues reported that Fyn phosphorylates the GABA(A)R gamma2 subunit on Tyr365 and Tyr367. In particular, phosphorylation on Tyr367 seems to be important for the modulation of GABA(A)R and consequently the inhibition of synaptic functions in mammalian brain [159].

Interactions of Fyn with other proteins in CNS have recently been reported, such as p250GAP, a RhoGAP protein expressed predominantly in the brain [160, 161], and RPTPrho, exclusively expressed in CNS [162, 163].

The description of the role of Fyn in phosphorylating tau and other specific proteins in CNS has increased the interest on this TK as a potential therapeutic target for neurodegenerative diseases.

7 Fyn in AD

Fyn hyperactivation/deregulation is involved in neurodegenerative disorders (tauopathies) characterised by alterations of tau, a microtubule-associated protein that is the main component of NFTs in AD. As previously described, the most important hallmarks of AD are the presence of SP and NFT. Plaques result from the self-aggregation of Aβ into insoluble fibrils, which represent the endpoint of an aggregation process generating a broad range of complexes of different molecular weights and solubility [164, 165]. NFTs are located inside the neurons, and their formation is a consequence of pathological post-translational modifications of tau [166, 167]. Indeed, NFTs are composed of highly phosphorylated forms of tau. The mechanisms leading to AD are not completely understood, but recent evidence suggests that alterations in Fyn kinase contribute to AD pathogenesis.

The initial evidence that Fyn may be implicated in some ways in the development of AD was reported in 1993 [168]. Shirazi and Wood found that a subset of neurons from AD brains exhibited a strong Fyn immunoreactivity compared to controls. The possible connection between Fyn and AD led to numerous studies in the past decades, focusing mainly on the effects of Aβ on TKs in cell culture. An early study suggested that treatment of cells with $A\beta_{25-35}$ and $A\beta_{1-40}$ increased tyrosine phosphorylation in a dose-dependent manner, although the exact Aβ preparation of the mixtures was not well characterised [169]. Even though the majority of early studies in AD used fibrillary or mixed preparations of Aβ, it is now widely believed that soluble intermediates of this peptide, termed Aβ oligomers (Aβo), are the most relevant assemblies in AD pathogenesis [170]. A critical early step in AD pathophysiology is the process by which Aβo interacts with the neuronal surface, and studies of this interaction have further implicated Fyn in AD pathophysiology. In a unique unbiased screen for cell surface proteins, Aβ was found to bind to the cellular prion protein (PrP^C) with high affinity and in a specific manner for the oligomeric form, with little or no affinity for fibrillary or monomeric Aβ peptide [171–174].

Once Aβo forms and binds to PrP^C at the cell surface, changes in neuronal biochemistry occur. Both PrP^C and Fyn were found to be enriched in the PSD fraction upon exposure of cultured mouse hippocampal neurons to Aβo, activating Fyn by phosphorylation of its Tyr420 residue [175]. Brain extracts from patients with AD also stimulate Fyn activation in mouse cortical cultures, while control brain extracts do not, further suggesting that the Aβo-PrP^C-Fyn pathway is relevant in human disease [175]. The connection between Aβo-PrP^C complexes at the cell surface and intracellular Fyn has recently been shown to require the metabotropic glutamate receptor mGluR5 [176]. This transmembrane protein has a physiological role in linking extracellular glutamate levels to calcium mobilisation, to protein translation in dendrites and to TK signalling. In neurons and in mice, mGluR5 is required for downstream neuronal dysfunction caused by Aβ. Extracellular Aβ oligomers thus trigger neuronal signal transduction from PrP^C to mGluR5 to Fyn kinase.

7.1 *Implications of Fyn-Tau Interaction in AD Development*

In addition to its role in Aβ signalling, Fyn is also involved in tau phosphorylation. Tau may play a critical role in mediating downstream neurodegeneration in AD, and a report showed that a 50% reduction in tau expression reversed cognitive impairments in a transgenic AD model [177]. Similarly to Aβo, studies on tau have mechanistically implicated Fyn in AD. Fyn physically associates with tau, via its SH3 domain, and phosphorylates tyrosine residues, including Tyr18, near the amino terminus [119, 178–180]. The phosphorylation of Tyr18 has been found in neurofibrillary tangles in human AD brain, suggesting a possible clinical relevance. Activation of Fyn by the Aβo-PrP^C complex was also demonstrated to lead to downstream tau phosphorylation [181].

Under normal conditions in neurons, tau is primarily located in axons but can also interact with Fyn in dendrites. The phosphorylation of NMDAR by Fyn was reported to stabilise the interaction between NMDAR and PSD95, a protein recruited into the NMDA receptor cluster [182]. This stabilisation enhances signalling by the neurotransmitter glutamate which strengthens Aβ toxicity. During AD development, the redistribution of hyperphosphorylated forms of tau from axons to dendrites may increase the tau-dependent storing of Fyn, leading to an abnormal NMDAR phosphorylation and consequently to toxic effects on neurons [183].

Fyn is not only involved in the development of AD by its interactions with NMDAR, tau and Aβ, but Fyn also binds 5-HT6R, widely distributed in the limbic region. 5-HT6R binds the Fyn SH3 domain via its C-terminal region, and activation of ERK1/2, a downstream target of 5-HT6R, is regulated in a Fyn-dependent manner [184].

8 Fyn Inhibitors for AD Treatment

Taken together, the preclinical evidence strongly supports that Fyn overexpression or hyperactivation is involved in brain pathologies, in particular AD. For this reason, the search for new molecular scaffolds able to inhibit Fyn activity represents an expanding field of study. Most Fyn inhibitors reported so far are ATP-competitive inhibitors, targeting the catalytic site of the enzyme. Unfortunately, because of the high sequence homology among SFKs, no specific Fyn inhibitor has been identified to date: all the reported compounds are also active against the other members of the family and sometimes against other TKs.

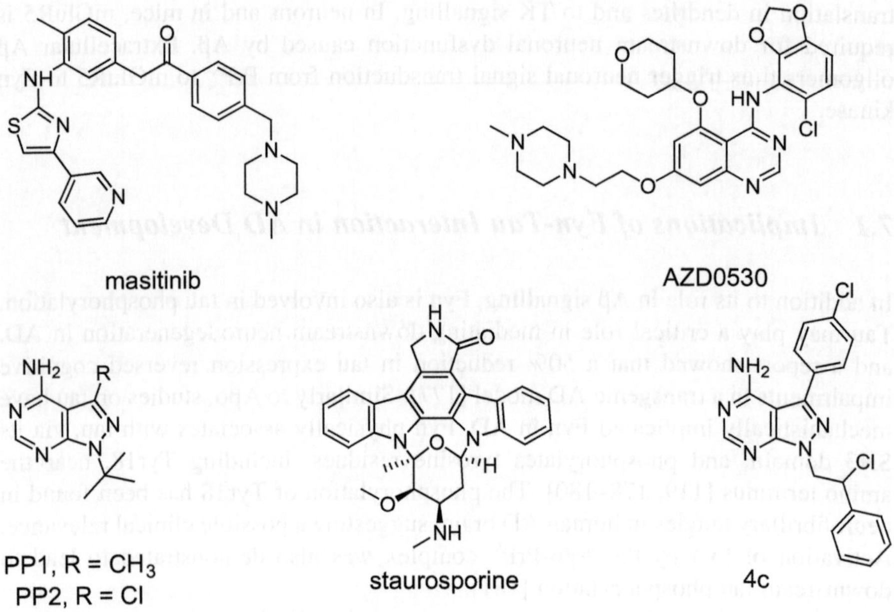

Fig. 5 Molecular structures of Fyn kinase inhibitors

8.1 Masitinib

Masitinib mesylate, also known as AB1010 (Fig. 5), is a phenylaminothiazole-type selective inhibitor targeting the ATP binding site of c-Kit receptor (c-KitR). Masitinib was originally approved for the treatment of mast cell tumours (MCT) in dogs [185, 186] and is currently under evaluation for the treatment of human cancers [187, 188]. In human medicine, masitinib is also considered a potential treatment for inflammatory diseases, including rheumatoid arthritis (RA), inflammatory bowel diseases and mastocytosis [119, 168, 189, 190].

In recent years, a new potential application of masitinib in neurodegenerative disease has emerged.

Even though masitinib is known as a selective inhibitor of c-Kit, it is also capable of blocking the SFK Fyn with a potency in the nanomolar range. As Fyn is involved in tau phosphorylation, it is possible that masitinib could prevent NFT formation. A 24-week randomised, placebo-controlled, phase II study was performed in patients with mild-to-moderate AD, receiving masitinib as an adjunct to a cholinesterase inhibitor and/or memantine [191]. Patients were randomly assigned to receive masitinib or placebo, administrated twice per day. A lower rate of clinically relevant cognitive decline was observed after 12 and 24 weeks with masitinib adjunct treatment compared with placebo. The trial provided evidence that masitinib could represent an innovative treatment for AD.

8.2 *Saracatinib (AZD0530)*

In 2006, AstraZeneca reported the discovery of AZD0530 (Fig. 5), also known as saracatinib, a small-molecule inhibitor of SFKs (Src, Fyn, Lyn and Yes) with a 2–10 nM potency [192]. At 10- to 100-fold higher concentrations, saracatinib was also active against Abl, but no detectable activity was found for other kinases, including c-Kit, Csk and PDGFR. Because of its selectivity against Fyn and SFKs, saracatinib was developed at first as therapy for solid tumours [192]. Clinical trials demonstrated that saracatinib is well tolerated and has a good oral bioavailability, but unfortunately a phase II study showed that it could not be used as a single agent in specific oncological indications because of its limited benefits [193–197].

In 2013, researchers from the Yale University School of Medicine launched a phase Ib study of saracatinib in patients with AD. The goal was to establish the safety, tolerability and CNS penetration of the drug. The study represented one of the prime success stories from the *Discovering New Therapeutics Uses for Existing Molecules* programme, established by the US National Institutes of Health in 2012. Typically in the drug development process, the passage from the discovery of a lead compound to the entry in a phase IIa trial takes at least one decade, but in the case of saracatinib, the research team completed all preclinical and clinical safety studies in 18 months. The multicentre phase IIa study is still recruiting participants, and the estimated primary completion date is December 2016 [ClinicalTrial.gov Identifier: NCT02167256].

9 Rational Design of Pyrazolo[3,4-D]pyrimidines as Fyn Kinase Inhibitors Active on AD Models

The pyrazolo[3,4-d]pyrimidines PP1 and PP2 (Fig. 5) were reported by Pfizer in 1996 as the first inhibitors of Lck and Fyn, with IC_{50} values ranging from 4 to 6 nM. Further studies indicated that PP1 is also active against Src and Hck kinase [198]. PP1 and PP2 are ATP-competitive inhibitors that bind the catalytic site of the enzyme, with the pyrazolopyrimidine core taking the place of the adenine ring of ATP. A large amount of biological data has been reported so far for PP1 and PP2, which are commonly used as reference compounds for newly synthesised derivatives or as pharmacological tools in the investigations on SFKs [199]. Even if biopharmaceutical properties of PP1 and PP2 appeared to limit their use as

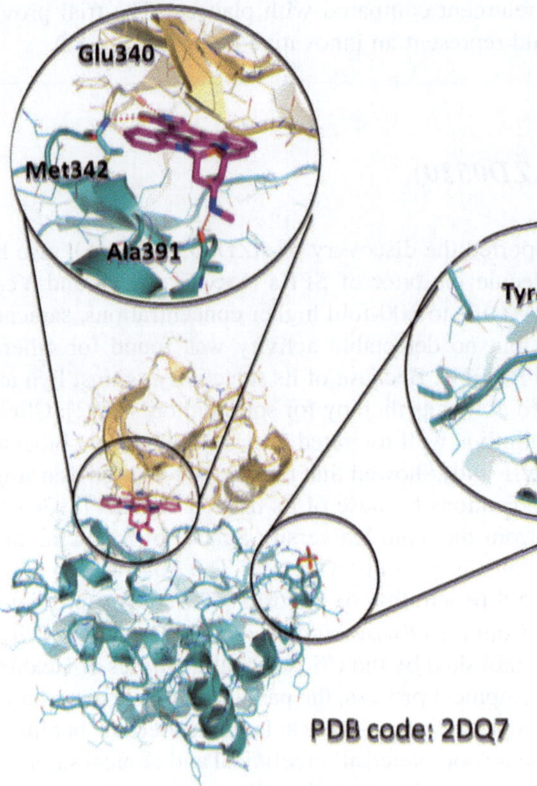

Fig. 6 X-ray crystal structure of the catalytic domain of Fyn (PDB code: 2DQ7). The N-terminal (small) lobe (*golden*) and the C-terminal (large) lobe (*cyano*) of the catalytic domain are highlighted. The C-lobe contains the activation loop with its Tyr417. Between the two lobes, the ATP-binding site is occupied by the staurosporine (*magenta*)

therapeutics, their discovery represents a significant advance in the use of TK inhibitors in the study of different cell functions.

Recently we reported the design, synthesis and biological evaluation of new pyrazolo[3,4-d]pyrimidine (PyrPyr) derivatives as selective inhibitors of Fyn kinase [200]. Selected compounds have been evaluated as anticancer agents, but surprisingly some of them were also found to be active in AD models. In particular, one molecule **4c** (Fig. 5) showed good potency in SH-SY5Y cell lines differentiated into neurons and treated with $A\beta_{42}$ oligomers/protofibrils in order to induce AD-like neurotoxicity [201, 202]. At first, **4c** was identified as a possible inhibitor of Fyn by applying a combination of molecular docking and molecular dynamics (MD) simulation protocols to an in-house library of pyrazolo[3,4-d]pyrimidines (consisting of ~300 structurally characterised compounds with >98% purity). The 2DQ7 [203] X-ray structure of the enzyme bound by staurosporine (Fig. 6) was retrieved from the Protein Data Bank (PDB) and used for computational studies.

Table 2 Molecular structures and enzymatic K_i values against Fyn of compounds **3a–p** and **4a**

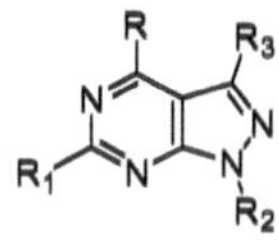

Cpd	R	R1	R2	R3	Fyn (K_i) (μM)
3a	$NHCH_2CH_2C_6H_5$	SC_4H_9	CH_2–CHCl–C_6H_5	H	NA
3b	NHC_6H_4-*m*Cl	SC_4H_9	CH_2–CHCl–C_6H_5	H	1.4
3c	$NHCH_2CH_2C_6H_5$	S-ciclopentyl	CH_2–CHCl–C_6H_5	H	NA
3d	NHC_6H_4-*m*F	S-ciclopentyl	CH_2–CHCl–C_6H_5	H	NA
3e	NHC_6H_4-*m*Cl	S-ciclopentyl	CH_2–CHCl–C_6H_5	H	NA
3f	NHC_3H_7	SCH_2CH_2-4morpholino	CH_2–CHCl–C_6H_5	H	8.15
3g	$NHCH_2C_6H_4$-*p*F	SCH_2CH_2-4morpholino	CH_2–CHCl–C_6H_5	H	15.5
3h	$NHCH_2CH_2C_6H_5$	SCH_2CH_2-4morpholino	CH_2–CHCl–C_6H_5	H	NA
3i	NHC_6H_5	SCH_2CH_2-4morpholino	CH_2–CHCl–C_6H_5	H	7.5
3j	$NHCH_2C_6H_5$	S-CH(CH_3)C_2H_5	CH_2–CHCl–C_6H_5	H	NA
3k	$NHCH_2CH_2C_6H_5$	S-CH(CH_3)C_2H_5	CH_2–CHCl–C_6H_5	H	NA
3l	NHC_6H_4-*m*Cl	S-CH(CH_3)C_2H_5	CH_2–CHCl–C_6H_5	H	12.5
3m	$NHCH_2CH_2C_6H_5$-*m*Cl	SCH_3	CH=CH–C_6H_5	H	16
3o	$NHCH_2C_6H_5$	$CH_2C_6H_5$	CH_2–CHCl–C_6H_5	H	13
3p	$NHCH_2C_6H_5$	H	CH_2–CHCl–C_6H_5	H	3.5
4a	NH_2	H	CH_2–CHCl–C_6H_5	C_6H_5	0.9

From a structural point of view, PyrPyr compounds are related to the Src inhibitors PP1 and PP2 but bear a different substitution pattern on the pyrazolo [3,4-d]pyrimidine nucleus. According to the predicted score values and the binding modes observed, 17 compounds were selected to be tested as Fyn inhibitors (Table 2). Among them, some derivatives (**3a–i**, **3p**) had already been reported by us as Bcr-Abl/Src inhibitors [204–207].

Three different binding modes emerged from docking studies. Compounds **3a–3o**, C3 unsubstituted and functionalised at positions 1, 4 and 6, showed an interaction network similar to that previously found within the ATP-binding site of c-Src (BM-A, Fig. 7a) [208]. The pyrazolo[3,4-d]pyrimidine core occupies the ATP adenine region, and the C4 substituent establishes hydrophobic interactions with Leu329, Phe409, Ile340, Phe311, Met318 and Leu411 (hydrophobic region I, HRI), while the N1 side chain is involved in van der Waals contacts with Tyr344, Met345 and Leu277 (hydrophobic region II, HRII). The C6 substituent is directed towards the external region. Two hydrogen bond interactions were also found, one involving the C4 amino group and the side chain of Thr342 and the other between the N2 of the pyrazolo[3,4-d]pyrimidine nucleus and the backbone NH group of Met345.

Conversely, the binding mode B (BM-B, Fig. 7b) was identified for compound **4a**, which contains a substituent at the C3 position while being unsubstituted at C6.

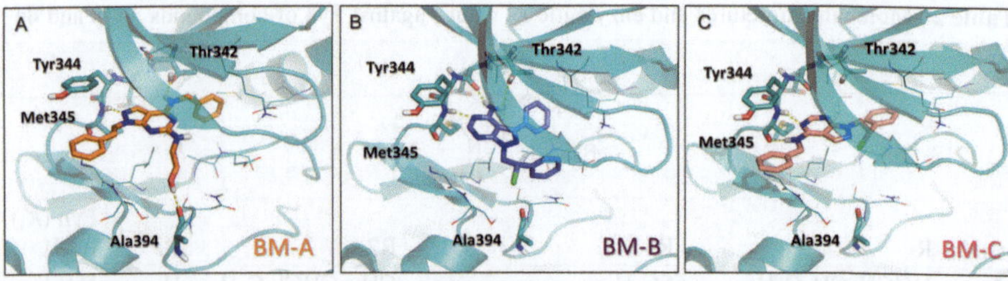

Fig. 7 3D graphical representation of predicted binding modes of pyrazolo[3,4-*d*]pyrimidines **3a–p** and **4a**. Three binding modes have been predicted, namely, BM-A (**A**), BM-B (7**B**) and BM-C (**C**). Compounds **3n** (*orange*), **4a** (*violet*) and **3p** (*pink*) are visualised

4a is a close analogue of PP1 and PP2, the first reported Fyn inhibitors [198]. Accordingly, its binding mode is consistent with that previously identified for PP1 and PP2 by crystallographic experiments [209, 210]. In detail, the exocyclic amine makes an H-bond with the carbonyl backbone of Glu343, while the nitrogen at N5 position acts as an H-bond acceptor in the interaction with Met345. The C3 phenyl group binds to the residues of hydrophobic region I through van der Waals interactions, while the N1 group is solvent exposed. Finally, compound **3p**, which is unsubstituted at both C3 and C6 positions, showed the binding mode C visualised in Fig. 7c. According to our prediction, the ligand binds the Met345 of the hinge region through two hydrogen bonds by means of the atom N5 and the exocyclic amino group at C4. Moreover, the N1 side chain is located within the hydrophobic pocket I formed by residues Leu329, Phe409, Ile340, Phe311, Met318 and Leu411, while the substituent at C4 interacts with Tyr344 and Leu277 of hydrophobic region II. Finally, biological investigation on isolated Fyn kinase (Table 2) revealed that **3b**, **3n**, **3p** and **4a** were the most active compounds, being endowed with K_i values of 1.4 μM, 1.15 μM, 3.5 μM and 0.9 μM, respectively. Derivatives **3f**, **3g**, **3i**, **3l**, **3m** and **3o** exhibited moderate activity with K_i ranging from 7.5 to 16 μM. No activity was detected for the remaining compounds.

Based on this finding, a hit-to-lead optimisation campaign was then initiated on compound **4a** to improve its potency. Nine analogues were synthesised (**4b–e**, **5a–e**) (Table 3).

The C3 moiety was differently functionalised, while minor modifications were carried out on the 2-chloro-2-phenylethyl N1 side chain by replacing or removing the chlorine atom. Furthermore, with the aim of exploring the effect caused by the introduction of a S-methyl moiety at the C6 position of the C3-phenyl-substituted pyrazolo[3,4-d]pyrimidine core, derivatives **6a** and **6b**, previously synthesised by our group [211], were added to this set of compounds. The 11 compounds were tested in a cell-free assay to evaluate their affinity towards isolated Fyn kinase.

As expected, derivatives **4c** and **4d** showed potent in vitro inhibitory effect against Fyn, with K_i values in the nanomolar range (70 nM and 95 nM, respectively). These potencies were most likely evoked due to the contribution of a chlorine or methyl substituent in para position of the C3 phenyl ring (compare **4c**

Table 3 Structure and activity against Fyn of compounds **4b–e**, **5a–e** and **6a–b**

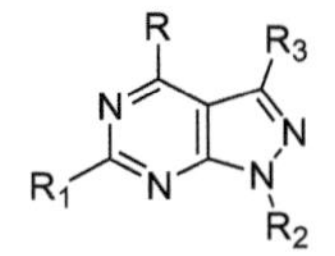

Cpd	R	R1	R2	R3	Fyn (K_i) (μM)
4b	NH_2	H	CH_2–CHCl–C_6H_5	C_6H_4–*p*F	0.36
4c	NH_2	H	CH_2–CHCl–C_6H_5	C_6H_4–*p*Cl	0.07
4d	NH_2	H	CH_2–CHCl–C_6H_5	C_6H_4–*p*CH_3	0.095
4e	NH_2	H	CH_2–CHCl–C_6H_5	C_6H_4–*p*OCH_3	1.485
5a	NH_2	H	CH_2–$CHCH_3$–C_6H_5	C_6H_5	3.35
5b	NH_2	H	CH_2–$CHCH_3$–C_6H_5	C_6H_4–*p*Cl	0.78
5c	NH_2	H	CH_2–$CHCH_3$–C_6H_5	C_6H_4–*p*CH_3	0.995
5d	NH_2	H	CH_2–$CHCH_3$–C_6H_5	C_6H_4–*p*$COCH_3$	>10
5e	NH_2	H	CH_2–$CHCH_3$–C_6H_5	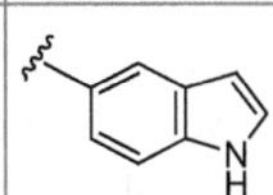	1.625
6a	$NHCH_3$	SCH_3	$CH_2CH_2C_6H_5$	C_6H_5	>10
6b	NH_2	SCH_3	$CH_2CH(OCH_3)$–C_6H_5	C_6H_5	>10

and **4d** with **4a**). Interesting activities were also found for compounds **4b**, **5b** and **5c** which exhibited submicromolar affinities (0.36 μM, 0.78 μM and 0.995 μM, respectively). As a general trend, the substitution of chlorine by methyl in the N1 side chain led to a considerable reduction of the affinity with about tenfold decreased activities (compare **5a** with **4a**, **5b** with **4c** and **5c** with **4d**). All studied compounds adopted a binding mode in line with that of **4a** with the exception of the inactive derivatives **5d** and **6a,b**.

In order to assess the stability of the complexes and to gain further insight into the ligand-protein interactions, molecular dynamics (MD) simulations were carried out starting from the geometry obtained by docking. The complexes between Fyn and derivatives **4a**, **4b**, **4c** and **4d** were selected as the training set. Each simulation was carried out for a total of 12 ns. Molecular mechanics generalised born surface area (MM-GBSA) approaches were then applied to estimate the free energy of binding between protein and ligands. The computed values are reported in Table 4. As expected, **4c**, the most active compound against Fyn kinase, showed the lowest favourable binding energy (-40.51 ± 2.63 kcal/mol).

Next, to determine the details of the interactions which play a pivotal role in the ligand-protein recognition process, a post-processing analysis was conducted for the four complexes. H-bonding interactions between ligands and enzyme as well as between the main residues of the active site were determined and are listed in Table 5 together with their occupancy in the investigated time period.

The analysis reveals that the H-bond which occurs between the nitrogen at position 5 (N5) of each ligand and the NH backbone of Met345 is weaker

Table 4 Calculated binding free energies of complexes between Fyn and compounds **4a–d**

Cpd	ΔG_{bind} MMGBSA (kcal/mol)
4a	−35.07 ± 2.25
4b	−38.21 ± 2.75
4c	−40.51 ± 2.63
4d	−38.36 ± 3.26

Table 5 Hydrogen bonds occupancy calculated from MD for **4a–d** ligands in complex with Fyn

Cpd	Glu343(O) N4	Met345(NH) N5	Thr342(OGH) Ala297(O)	Thr342(OGH) Glu343(O)	Thr342(NH) Ala297(O)
4a	67.4 %	33.3 %	54 %	5.7 %	88.3 %
4b	62 %	32.9 %	20 %	24.18 %	85.5 %
4c	76 %	32.8 %	53.2 %	–	85.7 %
4d	77.1 %	30.6 %	51.5 %	–	89.4 %

(occupancy of about 30%) than the one between the carbonyl backbone of Glu343 and the exocyclic amine (N4) (occupancy higher than 60%). As a consequence of this H-bond, the intermolecular interaction between the Thr342 side chain and the carbonyl backbone of Glu343 is weakened. This effect is particularly pronounced for the most active inhibitors **4c** and **4d**, for which the Thr342-Glu343 polar contact is totally lost. Furthermore, no interaction was detected between N4 and the Thr342 side chain, and this could be due to the fact that the "gatekeeper" residue is stabilised in its position by favourable interactions with Ala297 (Table 5). Such a network of interactions contributes to the stabilisation of the binding of our PyrPyr together with the enzyme. The global effect of stabilisation was more pronounced for the most active inhibitors, in line with the trend observed for the activity against Fyn.

9.1 Selectivity Against a Human Kinase Panel

To assess its specificity against Fyn, compound **4c** was tested in a panel of kinases including other Src family members (Hck, Blk, Fgr, Fyn, Src, Lck, Lyn and Yes), tyrosine kinases (Abl, EGFR, IGF-1R, JAK2, PDGFR, KDR) as well as some serine-threonine kinases. The percentage of enzymatic activity was measured for each kinase by using **4c** at 10 μM (Millipore); results are listed in Table 6.

Remarkably, **4c** proved to be more efficient against SFKs than towards the other investigated kinases, confirming such a compound as a useful probe to study Src family function. Furthermore, high activity against Abl was also detected, as expected for the high structural similarity between Abl and the SFKs. Notably, **4c** was not able to significantly inhibit any of the serine-threonine kinases tested (Pim-1, mTOR, JNK, CDK5, CHK1) and also was not active towards the serine-threonine kinases DYRK1a and GSK3-β that were reported as implicated in AD [212].

Table 6 Kinase profiling of compounds **4c** by Millipore

	Residual activity (%)
Fyn (h)	1
Abl(h)	9
Abl (H396P) (h)	5
Abl (M351T)(h)	6
Abl (Q252H) (h)	3
Abl(T315I)(h)	97
Abl(Y253F)(h)	2
Blk(h)	19
CDK5/p25(h)	86
CDK5/p35(h)	99
CHK1(h)	97
cKit(h)	36
cSRC(h)	17
EGFR(h)	47
Fgr(h)	4
GSK3β(h)	66
Hck(h)	7
Hck(h) activated	2
IGF-1R(h)	71
JAK2(h)	118
JNK3(h)	53
KDR(h)	31
Lck(h)	2
Lck(h) activated	9
Lyn(h)	15
mTOR(h)	47
PDGFRβ(h)	51
Pim-1(h)	91
Src(1–530)(h)	2
Src(T341M)(h)	98
Yes(h)	5
DYRK1a	85

9.2 *Biological Evaluation in Alzheimer's Disease Model*

Since Fyn mediates the phosphorylation of tau on the Tyr18 residue and is therefore considered a promising therapeutic target for the treatment of AD, the most interesting compounds identified during in vitro inhibition assays, **4c** and **4d**, were evaluated for their ability to inhibit the Fyn-mediated phosphorylation of residue Tyr18 on tau in an AD model cell line. To this aim, neuroblastoma SH-SY5Y cells were differentiated to mature neurons with the administration of retinoic acid, followed by brain-derived neurotrophic factor, neuregulin β1, nerve growth factor and vitamin D3 treatment. Once differentiated, SH-SY5Y cells were treated with $A\beta_{42}$ oligomer/protofibril in order to induce AD-like neurotoxicity [201, 213]. Both compounds significantly affected $A\beta_{42}$-induced Tyr18-tau phosphorylation to a similar degree and in a dose-dependent manner [200]. Moreover, the inhibitory activity of **4c** and **4d** was constant over time, being effective up to 6 h after compound administration [200].

10 Conclusions

The accelerating pace of population ageing is producing an increased prevalence of those neurodegenerative diseases that appear in the late stages of life and that require persistent caregiving, for example, AD. Despite the important impact on public health costs that this provokes, current therapy does not stop neuronal degeneration and only ameliorates symptoms. Thus, there is a compelling need for the development of AD disease-modifying agents. Some of these molecules are in clinical trial studies, but discouraging results indicate that more efforts are needed. The increasing understanding of AD aetiology indicates that tau-phosphorylating kinases are strongly involved in the disease and thus represent promising drug targets. However, the difficulties in the development of GSK3β inhibitors suggest that research should move towards other tau-phosphorylating kinases.

Fyn is responsible of the development and regulation of the physiological functions of the CNS, and its overexpression or hyperactivation is strongly correlated with tauopathies. Interestingly, in addition to tau hyperphosphorylation, several other cascades link Fyn to AD aetiology, making this kinase a very promising anti-AD drug target. Indeed, Fyn ATP-competitive inhibitors are in preclinical and clinical development with encouraging results. This might indeed be one of the drug development strategies to follow for the development of efficacious disease-modifying drugs for AD.

References

1. NIH National Institute of Aging. Alzheimer's Disease Fact Sheet. NIH Publication No. 15-6423, May 2015. Available at https://www.nia.nih.gov/alzheimers/publication/alzheimers-disease-fact-sheet
2. Berchtold NC, Cotman CW (1998) Evolution in the conceptualization of Dementia and Alzheimer's disease: Greco-Roman period to the 1960s. Neurobiol Aging 19:173–189
3. Stelzmann RA, Schnitzlein HN, Murtagh FR (1995) An English Translation of Alzheimer's 1907 Paper, "Über eine eigenartige Erlranliung der Hirnrinde". Clin Anat 8:429–431
4. Kidd M (1963) Paired helical filaments in electron microscopy of Alzheimer's disease. Nature 197:192–193
5. Alzheimer's Association (2016) http://www.alz.org/
6. World Health Organization (2015) http://www.who.int/mediacentre/factsheets/fs404/en/

7. Qui C (2009) Epidemiology of Alzheimer's disease: occurrence, determinants, and strategies toward intervention dialogues. Clin Neurosci 11:111–128
8. Alzheimer's Association (2016) Alzheimer's disease facts and figures. Alzheimers Dement 12:459–509
9. Hebert LE, Weuve J, Scherr PA et al (2013) Alzheimer disease in the United States (2010–2050) estimated using the 2010 census. Neurology 80:1778–1783
10. McKhann G, Drachman D, Folstein M et al (1984) Clinical diagnosis of Alzheimer's disease: report of the NINCDS-ADRDA Work Group under the auspices of Department of Health and Human Services Task Force on Alzheimer's Disease. Neurology 34:939–944
11. Jack CR, Albert MS, Knopman DS et al (2011) Introduction to the recommendations from the National Institute on Aging-Alzheimer's Association workgroups on diagnostic guidelines for Alzheimer's disease. Alzheimers Dement 7:257–262
12. McKhann GM, Knopman DS, Chertkow H et al (2011) The diagnosis of dementia due to Alzheimer's disease: recommendations from the National Institute on Aging-Alzheimer's Association workgroups on diagnostic guidelines for Alzheimer's disease. Alzheimers Dement 7:263–269
13. Albert MS, DeKosky ST, Dickson D et al (2011) The diagnosis of mild cognitive impairment due to Alzheimer's disease: recommendations from the National Institute on Aging-Alzheimer's Association workgroups on diagnostic guidelines for Alzheimer's disease. Alzheimers Dement 7:270–279
14. Sperling RA, Aisen PS, Becket LA et al (2011) Toward defining the preclinical stages of Alzheimer's disease: recommendations from the National Institute on Aging-Alzheimer's Association workgroups on diagnostic guidelines for Alzheimer's disease. Alzheimers Dement 7:280–292
15. Reitz C, Mayeux R (2014) Alzheimer disease: epidemiology, diagnostic criteria, risk factors and biomarkers. Biochem Pharmacol 88:640–651
16. Rosenberg RN (2000) The molecular and genetic basis of AD: the end of the beginning: the 2000 Wartenberg lecture. Neurology 54:2045–2054
17. Crews L, Masliah E (2010) Molecular mechanisms of neurodegeneration in Alzheimer's disease. Hum Mol Genet 19:R12–R20
18. Terry RD, Masliah E, Salmon DP et al (1991) Physical basis of cognitive alterations in Alzheimer's disease: synapse loss is the major correlate of cognitive impairment. Ann Neurol 30:572–580
19. DeKosky S, Scheff S (1990) Synapse loss in frontal cortex biopsies in Alzheimer's disease: correlation with cognitive severity. Ann Neurol 27:457–464
20. Monczor M (2005) Diagnosis and treatment of Alzheimer's disease. Curr Med Chem 5:5–13
21. Fountoulakis M, Kossida S (2006) Proteomics-driven progress in neurodegeneration research. Electrophoresis 27:1556–1573
22. Akiyama H, Barger S, Barnum S et al (2000) Inflammation and Alzheimer's disease. Neurobiol Aging 21:383–421
23. Wyss-Coray T (2006) Inflammation in Alzheimer disease: driving force, bystander or beneficial response? Nat Med 12:1005–1015
24. McGeer PL, Schulzer M, McGeer EG (1996) Arthritis and anti-inflammatory agents as possible protective factors for Alzheimer's disease: a review of 17 epidemiologic studies. Neurology 47:425–432
25. Bartus RT, Dean RL, Beer B et al (1982) The cholinergic hypothesis of geriatric memory dysfunction. Science 217:408–417
26. Perry EK (1986) The cholinergic hypothesis-ten years on. Br Med Bull 42:63–69
27. Contestabile A (2011) The history of the cholinergic hypothesis. Behav Brain Res 221:334–340
28. Cacabelos R, Takeda M, Winbland B (1999) The glutamatergic system and neurodegeneration in dementia: preventive strategies in Alzheimer's disease. Int J Geriatr Psychiatry 14:3–47

29. Selkoe DJ (1991) The molecular pathology of Alzheimer's disease. Neuron 6:487–498
30. Hardy JA, Higgings GA (1992) Alzheimer's disease: the amyloid cascade hypothesis. Science 256:184–185
31. Karran E, De Strooper B (2016) The amyloid cascade hypothesis: are we poised for success or failure? J Neurochem. doi:10.1111/jnc.13632 (published ahead of print, June 2016)
32. Morishima-Kawashima M, Ihara Y (2002) Alzheimer's disease: β-amyloid protein and tau. Neurosci Res 70:392–401
33. Querfurth HW, LaFerla FM (2010) Alzheimer's disease. N Engl J Med 362:329–344
34. Johnson RD, Steel DG, Gafni A (2014) Structural evolution and membrane interactions of Alzheimer's amyloid-β peptide oligomers: new knowledge from single-molecule fluorescence studies. Protein Sci 23:869–883
35. Iqbal K, Liu F, Gong CX et al (2010) Tau in Alzheimer disease and related tauopathies. Curr Alzheimer Res 7:656–664
36. Gendron TF, Petrucelli L (2009) The role of tau in neurodegeneration. Mol Neurodegener 4:13
37. Martin L, Latypova X, Wilson CM et al (2013) Tau protein kinases: involvement in Alzheimer's disease. Ageing Res Rev 12:289–309
38. Drewes G, Trinczek B, Illenberger S et al (1995) Microtubule-associated protein/microtubule affinity-regulating kinase (p110mark). A novel protein kinase that regulates tau-microtubule interactions and dynamic instability by phosphorylation at the Alzheimer-specific site serine 262. J Biol Chem 270:7679–7688
39. Sengupta A, Kabat J, Novak M et al (1998) Phosphorylation of tau at both Thr231 and Ser262 is required for maximal inhibition of its binding to microtubules. Arch Biochem Biophys 357:299–309
40. Sergeant N, Delacourte A, Buée L (2005) Tau protein as a differential biomarker of tauopathies. Biochim Biophys Acta 1739:179–197
41. Yiannopoulou KG, Papageorgiou SG (2013) Current and future treatments for Alzheimer's disease. Ther Adv Neurol Disord 6:19–33
42. Cheng X, Zhang L, Lian YJ (2015) Molecular targets in Alzheimer's disease: from pathogenesis to therapeutics. Biomed Res Int 2015:760758
43. Birks J, Harvey RJ (2006) Donepezil for dementia due to Alzheimer's disease. Cochrane Database Syst Rev 1, CD001190
44. Scott LJ, Goa KL (2000) Galantamine: a review of its use in Alzheimer's disease. Drugs 60:1095–1122
45. Reisberg B, Doody R, Stöffler A et al (2003) Memantine in moderate-to-severe Alzheimer's disease. N Engl J Med 348:1333–1341
46. Winblad B, Grossberg G, Frölich L et al (2007) A 6-month, double-blind, placebo-controlled study of the first skin patch for Alzheimer disease. Neurology 69:S14–S22
47. Atri A, Hendrix SB, Pejović V et al (2015) Cumulative, additive benefits of memantine-donepezil combination over component monotherapies in moderate to severe Alzheimer's dementia: a pooled area under the curve analysis. Alzheimers Res Ther 7:28
48. Hansen RA, Gartlehner G, Webb AP et al (2008) Efficacy and safety of donepezil, galantamine, and rivastigmine for the treatment of Alzheimer's disease: a systematic review and meta-analysis. Clin Interv Aging 3:211–225
49. Maidment ID, Fox CG, Boustani M et al (2008) Efficacy of memantine on behavioural and psychological symptoms related to dementia: a systematic meta-analysis. Ann Pharmacother 42:32–38
50. Lee G, Neve RL, Kosik KS (1989) The microtubule binding domain of tau protein. Neuron 2:1615–1624
51. LaPointe NE, Morfini G, Pigino G et al (2009) The amino terminus of tau inhibits kinesin-dependent axonal transport: implications for filament toxicity. J Neurosci Res 87:440–451
52. Arriagada PV, Growdon JH, Hedley-Whyte ET et al (1992) Neurofibrillary tangles but not senile plaques parallel duration and severity of Alzheimer's disease. Neurology 42:631–639

53. Embi N, Rylatt DB, Cohen P (1980) Glycogen synthase kinase-3 from rabbit skeletal muscle. Separation from cyclic-AMP-dependent protein kinase and phosphorylase kinase. Eur J Biochem 107:519–527
54. Eldar-Finkelman H, Martinez A (2011) GSK-3 inhibitors: preclinical and clinical focus on CNS. Front Mol Neurosci 4:32
55. Hooper C, Killick R, Lovestone S (2008) The GSK3 hypothesis of Alzheimer's disease. J Neurochem 104:1433–1439
56. Rayasam GV, Tulasi VK, Sodhi R et al (2009) Glycogen synthase kinase 3: more than a name sake. Br J Pharmacol 156:885–898
57. Beurel E, Michalek SM, Jope RS (2010) Innate and adaptive immune responses regulated by glycogen synthase kinase-3 (GSK3). Trends Immunol 31:24–31
58. Woodgett JR (1990) Molecular cloning and expression of glycogen synthase kinase-3/factor A. EMBO J 9:2431–2438
59. Mukai F, Ishiguro K, Sano Y et al (2002) Alternative splicing isoform of tau protein kinase 1/glycogen synthase kinase 3beta. J Neurochem 81:1073–1083
60. Hanger DP, Anderton BH, Noble W (2009) Tau phosphorylation: the therapeutic challenge for neurodegenerative disease. Trends Mol Med 15:112–119
61. Lucas JJ, Hernández F, Gómez-Ramos P et al (2001) Decreased nuclear beta-catenin, tau hyperphosphorylation and neurodegeneration in GSK-3beta conditional transgenic mice. EMBO J 20:27–39
62. Hoshi M, Takashima A, Noguchi K et al (1996) Regulation of mitochondrial pyruvate dehydrogenase activity by tau protein kinase I/glycogen synthase kinase 3beta in brain. Proc Natl Acad Sci U S A 93:2719–2723
63. Gómez de Barreda E, Pérez M, Gómez Ramos P et al (2010) Tau-knockout mice show reduced GSK3-induced hippocampal degeneration and learning deficits. Neurobiol Dis 37:622–629
64. Yamaguchi H, Ishiguro K, Uchida T et al (1996) Preferential labelling of Alzheimer neurofibrillary tangles with antisera for tau protein kinase (TPK) I/glycogen synthase kinase-3 beta and cyclin-dependent kinase 5, a component of TPK II. Acta Neuropathol 92:232–241
65. Pei JJ, Tanaka T, Tung YC et al (1997) Distribution, levels, and activity of glycogen synthase kinase-3 in the Alzheimer disease brain. J Neuropathol 56:70–78
66. Leroy K, Yilmaz Z, Brion JP (2007) Increased level of active GSK-3beta in Alzheimer's disease and accumulation in argyrophilic grains and in neurons at different stages of neurofibrillary degeneration. Neuropathol Appl Neurobiol 33:43–55
67. Smith D (2003) Cdk5 in neuroskeletal dynamics. Neurosignals 12:239–251
68. Nikolic M, Dudek H, Kwon YT et al (1996) The cdk5/p35 kinase is essential for neurite outgrowth during neuronal differentiation. Genes Dev 10:816–825
69. Barclay JW, Aldea M, Craig TJ et al (2007) Regulation of the fusion pore conductance during exocytosis by cyclin-dependent kinase 5. J Biol Chem 279:41495–41503
70. Samuels BA, Hsueh YP, Shu T et al (2007) Cdk5 promotes synaptogenesis by regulating the subcellular distribution of the MAGUK family member CASK. Neuron 56:823–837
71. Humbert S, Dhavan R, Tsai L (2000) p39 activates cdk5 in neurons, and is associated with the actin cytoskeleton. J Cell Sci 113:975–983
72. Dhariwala FA, Rajadhyaksha MS (2008) An unusual member of the Cdk family: Cdk5. Cell Mol Neurobiol 28:351–369
73. Patrick GN, Zukerberg L, Nikolic M et al (1999) Conversion of p35 to p25 deregulates Cdk5 activity and promotes neurodegeneration. Nature 402:615–622
74. Noble W, Olm V, Takata K et al (2003) Cdk5 is a key factor in tau aggregation and tangle formation in vivo. Neuron 38:555–565
75. Sengupta A, Wu Q, Grundke-Iqbal I et al (1997) Potentiation of GSK-3-catalyzed Alzheimer-like phosphorylation of human tau by cdk5. Mol Cell Biochem 167:99–105

76. Cheung ZH, Gong K, Ip NY (2008) Cyclin-dependent kinase 5 supports neuronal survival through phosphorylation of Bcl-2. J Neurosci 28:4872–4877
77. Lopes JP, Oliveira CR, Agostinho P (2010) Neurodegeneration in Aβ-induced model of Alzheimer's disease: the role of Cdk5. Aging Cell 9:64–77
78. Atzori C, Ghetti B, Piva R et al (2001) Activation of the JNK/p38 pathway occurs in diseases characterized by tau protein pathology and is related to tau phosphorylation but not apoptosis. J Neuropathol Exp Neurol 60:1190–1197
79. Ferrer I, Blanco R, Carmona M et al (2001) Phosphorylated map kinase (ERK1, ERK2) expression is associated with early tau deposition in neurones and glial cells, but not with increased nuclear DNA vulnerability and cell death, in Alzheimer disease. Brain Pathol 11:144–158
80. Remy G, Risco AM, Iñesta-Vaquera FA et al (2010) Differential activation of p38MAPK isoforms by MKK6 and MKK3. Cell Signal 22:660–667
81. Sergeant N, Bretteville A, Hamdane M et al (2008) Biochemistry of tau in Alzheimer's disease and related neurological disorders. Expert Rev Proteomics 5:207–224
82. Thornton TM, Pedraza-Alva G, Deng B et al (2008) Phosphorylation by p38 MAPK as an alternative pathway for GSK3beta inactivation. Science 320:667–670
83. Yang Y, Zhu X, Chen Y et al (2007) p38 and JNK MAPK, but not ERK1/2 MAPK, play an important role in cholchicine-induced cortical neurons apoptosis. Eur J Pharmacol 576:26–33
84. Cuadrado A, Nebreda AR (2010) Mechanisms and functions of p38 MAPK signalling. Biochem J 429:403–417
85. Zhu X, Rottkamp CA, Boux H et al (2000) Activation of p38 links tau phosphorylation, oxidative stress, and cell cycle-related events in Alzheimer disease. J Neuropathol Exp Neurol 59:880–888
86. Munoz L, Ammit AJ (2010) Targeting p38 MAPK pathway for the treatment of Alzheimer's disease. Neuropharmacology 58:561–568
87. Guise S, Braguer D, Carles G et al (2001) Hyperphosphorylation of tau is mediated by ERK activation during anticancer drug-induced apoptosis in neuroblastoma cells. J Neurosci Res 63:257–267
88. Medina MG, Ledesma MD, Domínguez JE et al (2005) Tissue plasminogen activator mediates amyloid-induced neurotoxicity via Erk1/2 activation. EMBO J 24:1706–1716
89. Pei JJ, Braak H, An WL et al (2002) Up-regulation of mitogen-activated protein kinase ERK1/2 and MEK1/2 is associated with the progression of neurofibrillary degeneration in Alzheimer's disease. Brain Res Mol Brain Res 109:45–55
90. Wada T, Penninger JM (2004) Mitogen-activated protein kinases in apoptosis regulation. Oncogene 23:2838–2849
91. Zhu X, Raina AK, Rottkamp CA et al (2001) Activation and redistribution of c-jun-N-terminal kinase/stress activated protein kinase in degenerating neurons in Alzheimer's disease. J Neurochem 76:435–441
92. Shen C, Chen Y, Liu H et al (2008) Hydrogen peroxide promotes Abeta production through JNK-dependent activation of gamma-secretase. J Biol Chem 283:17721–17730
93. Houlden H, Johnson J, Gardner-Thorpe C et al (2007) Mutations in TTBK2, encoding a kinase implicated in tau phosphorylation, segregate with spinocerebellar ataxia type 11. Nat Genet 39:1434–1436
94. Sato S, Xu J, Okuyama S et al (2008) Spatial learning impairment, enhanced CD5/p35 activity, and downregulation of NMDA receptor expression in transgenic tau-tubulin kinase 1. J Neurosci 28:14511–14521
95. Li G, Yin H, Kuret J (2004) Casein kinase 1 delta phosphorylates tau and disrupts its binding to microtubules. J Biol Chem 279:15938–15945
96. Hanger DP, Byers HL, Wray S et al (2007) Novel phosphorylation sites in tau from Alzheimer brain support a role for casein kinase 1 in disease pathogenesis. J Biol Chem 282:23645–23654

97. Liu C, Li Y, Semenov M et al (2002) Control of beta-catenin phosphorylation/degradation by a dual-kinase mechanism. Cell 108:837–847
98. Ahmad KA, Wang G, Unger G et al (2008) Protein kinase CK2-a key suppressor of apoptosis. Adv Enzyme Regul 48:179–187
99. Knippschild U, Gocht A, Wolff S et al (2005) The casein kinase I family: participation in multiple cellular processes in eukaryotes. Cell Signal 17:675–689
100. Chauhan A, Chauhan VP, Murakami N (1993) Amyloid beta-protein stimulates casein kinase I and casein kinases II activities. Brain Res 629:47–52
101. Schwab C, DeMaggio AJ, Ghoshal N et al (2000) Casein kinase I delta is associated with pathological accumulation of tau in several neurodegenerative diseases. Neurobiol Aging 21:503–510
102. Yasojima K, Kuret J, DeMaggio AJ et al (2000) Casein kinase 1 delta mRNA is upregulated in Alzheimer disease brain. Brain Res 865:116–120
103. Ferrer I, Barrachina M, Puig B et al (2005) Constitutive Dyrk1A is abnormally expressed in Alzheimer disease, Down syndrome, Pick disease, and related transgenic models. Neurobiol Dis 20:392–400
104. Yang EJ, Ahn YS, Chung KC (2001) Protein kinase Dyrk1 activates cAMP response element-binding protein during neuronal differentiation in hippocampal progenitor cells. J Biol Chem 276:39819–39824
105. Chen H, Antonarakis SE (1997) Localisation of a human homologue of the Drosophila mnb and rat Dyrk genes to chromosome 21q22.2. Hum Genet 99:262–265
106. Masters CL, Simms G, Weinman NA et al (1985) Amyloid plaque core protein in Alzheimer disease and Down syndrome. Proc Natl Acad Sci U S A 82:4245–4249
107. Kimura R, Kamino K, Yamamoto M et al (2007) The DYRK1A gene, encoded by chromosome 21 Down syndrome critical region, bridges between beta-amyloid production and tau phosphorylation in Alzheimer disease. Hum Mol Genet 16:15–23
108. Dell'Acqua ML, Scott JD (1997) Protein kinase A anchoring. J Biol Chem 272:12881–12884
109. Tian Q, Zhang JX, Zhang Y et al (2009) Biphasic effects of forskolin on tau phosphorylation and spatial memory in rats. J Alzheimers Dis 17:631–642
110. Soderling TR (1999) The Ca-calmodulin-dependent protein kinase cascade. Trends Biochem Sci 24:232–236
111. Xie S, Jin N, Gu J et al (2016) O-GlcNAcylation of protein kinase A catalytic subunits enhances its activity: a mechanism linked to learning and memory deficits in Alzheimer's disease. Aging Cell 15:455–464
112. Kitagawa M, Mukai H, Shibata H et al (1995) Purification and characterization of a fatty acid-activated protein kinase (PKN) from rat testis. Biochem J 310:657–664
113. Taniguchi T, Kawamata T, Mukai H et al (2001) Phosphorylation of tau is regulated by PKN. J Biol Chem 276:10025–10031
114. Isagawa T, Mukai H, Oishi K et al (2000) Dual effects of PKNalpha and protein kinase C on phosphorylation of tau protein by glycogen synthase kinase-3beta. Biochem Biophys Res Commun 273:209–212
115. Grifith LC (2004) Regulation of calcium/calmodulin-dependent protein kinase II activation by intramolecular and intermolecular interactions. J Neurosci 24:8394–8398
116. Yamamoto H, Hiragami Y, Murayama M et al (2005) Phosphorylation of tau at serine 416 by Ca2+/calmodulin-dependent protein kinase II in neuronal soma in brain. J Neurochem 94:1438–1447
117. Wang YJ, Chen GH, Hu XY et al (2005) The expression of calcium/calmodulin-dependent protein kinase II-alpha in the hippocampus of patients with Alzheimer's disease and its links with AD-related pathology. Brain Res 1031:101–108
118. Tsukane M, Yamauchi T (2009) Ca2+/calmodulin-dependent protein kinase II mediates apoptosis of P19 cells expressing human tau during neural differentiation with retinoic acid treatment. J Enzyme Inhib Med Chem 24:365–371

119. Lee G, Thangavel R, Sharma VM et al (2004) Phosphorylation of tau by fyn: implications for Alzheimer's disease. J Neurosci 24:2304–2312
120. Derkinderen P, Scales TM, Hanger DP et al (2005) Tyrosine 394 is phosphorylated in Alzheimer's paired helical filament tau and in fetal tau with c-Abl as the candidate tyrosine kinase. J Neurosci 25:6584–6593
121. Vega IE, Cui L, Propst JA et al (2005) Increase in tau tyrosine phosphorylation correlates with the formation of tau aggregates. Brain Res 138:135–144
122. Lebouvier T, Scales TM, Williamson R et al (2009) The microtubule-associated tau is also phosphorylated on tyrosine. J Alzheimers Dis 18:1–9
123. Van Etten RA (1999) Cycling, stressed-out and nervous: cellular functions of c-Abl. Trends Cell Biol 9:179–186
124. Plattner R, Kadlec L, DeMali KA et al (1999) c-Abl is activated by growth factors and Src family kinases and has a role in the cellular response to PDGF. Genes Dev 13:2400–2411
125. Vázquez MC, Vargas LM, Inestrosa NC et al (2009) c-Abl modulates AICD dependent cellular responses: transcriptional induction and apoptosis. J Cell Physiol 220:136–143
126. Tell V, Hilgeroth A (2013) Recent developments of protein kinase inhibitors as potential AD therapeutics. Front Cell Neurosci 7:189
127. Godyń J, Jończyk J, Panek D et al (2016) Therapeutic strategies for Alzheimer's disease in clinical trials. Pharmacol Rep 68:127–138
128. Silva T, Reis J, Teixera J et al (2014) Alzheimer's disease, enzyme targets and drug discovery struggles: from natural products to drug prototypes. Ageing Res Rev 15:116–145
129. Kumar A, Singh A, Ekavali (2015) A review on Alzheimer's disease pathophysiology and its management: an update. Pharmacol Rep 67:195–203
130. Ellis L (2016) Alzheimer's disease: defusing the bomb by 2015. IN VIVO: The Business & Medicine Report 34:30–37
131. Geldenhuys WJ, Darvesh AS (2015) Pharmacotherapy of Alzheimer's disease: current and future trends. Expert Rev Neurother 15:3–5
132. Mairet-Coello G, Polleux F (2014) Involvement of 'stress–response' kinase pathways in Alzheimer's disease progression. Curr Opin Neurobiol 27:110–117
133. Krug M, Hilgeroth A (2008) Recent advances in the development of multi-kinase inhibitors. Mini Rev Med Chem 8:1312–1327
134. Zhang J, Yang PL, Gray NS (2009) Targeting cancer with small molecule kinase inhibitors. Nat Rev Cancer 9:28–29
135. Nygaard HB, van Dyck CH, Strittmatter SM (2014) Fyn kinase inhibition as a novel therapy for Alzheimer's disease. Alzheimers Res Ther 6:8
136. Summy JM, Gallick GE (2003) Src family kinases in tumor progression and metastasis. Cancer Metastasis Rev 22:337–358
137. Guarino M (2010) Src signaling in cancer invasion. J Cell Physiol 223:14–26
138. Summy JM, Gallick GE (2003) Treatment for advanced tumors: SRC reclaims center stage. Clin Cancer Res 12:1398–1401
139. Goel RK, Lukong KE (2016) Understanding the cellular roles of Fyn-related kinase (FRK): implications in cancer biology. Cancer Metastasis Rev 35:179–199
140. Thomas SM, Brugge JS (1997) Cellular functions regulated by Src family kinases. Annu Rev Cell Dev Biol 13:513–609
141. Goldsmith JF, Hall CG, Atkinson TP (2002) Identification of an alternatively spliced isoform of the fyn tyrosine kinase. Biochem Biophys Res Commun 298:501–504
142. Knighton DR, Zheng JH, Ten Eyck LF et al (1991) Crystal structure of the catalytic subunit of cyclic adenosine monophosphate-dependent protein kinase. Science 253:407–414
143. Songyang Z, Steven ES, Manas C et al (1993) SH2 domains recognize specific phosphopeptide sequences. Cell 72:767–778
144. Songyang Z, Shoelson SE, McGlade J et al (1994) Specific motifs recognized by the SH2 domains of Csk, 3BP2, fps/fes, GRB-2, HCP, SHC, Syk, and Vav. Mol Cell Biol 14:2777–2785

145. Cohen GB, Ren R, Baltimore D (1995) Modular binding domains in signal transduction proteins. Cell 80:237–248
146. Ren R, Mayer BJ, Cicchetti P et al (1993) Identification of a ten-amino acid proline-rich SH3 binding site. Science 259:1157–1161
147. Mayer BJ, Eck MJ (1995) SH3 domains. Minding your p's and q's. Curr Biol 5:364–367
148. zur Hausen JD, Burn P, Amrein KE (1997) Co-localization of Fyn with CD3 complex, CD45 or CD28 depends on different mechanisms. Eur J Immunol 27:2643–2649
149. Sigal CT, Zhou W, Buser CA et al (1994) Amino-terminal basic residues of Src mediate membrane binding through electrostatic interaction with acidic phospholipids. Proc Natl Acad Sci U S A 91:12253–12257
150. Alland L, Peseckis SM, Atherton RE et al (1994) Dual myristoylation and palmitoylation of Src family member p59fyn affects subcellular localization. J Biol Chem 269:16701–16705
151. Salmond RJ, Filby A, Qureshi I et al (2009) T-cell receptor proximal signaling via the Src-family kinases, Lck and Fyn, influences T-cell activation, differentiation and tolerance. Immunol Rev 228:9–22
152. Grant SG, O'Dell TJ, Karl KA et al (1992) Impaired long-term potentiation, spatial learning and hippocampal development in fyn mutant mice. Science 258:1903–1910
153. Osterhout DJ, Wolven A, Wolven A, Wolf RM et al (1999) Morphological differentiation of oligodendrocytes requires activation of Fyn tyrosine kinase. J Cell Biol 145:1209–1218
154. Macurek L, Dráberová L, Marková V et al (2008) Regulation of microtubule nucleation from membranes by complexes of membrane-bound gamma-tubulin with Fyn kinase and phosphoinositide 3-kinase. Biochem J 416:421–430
155. Tezuka T, Umemori H, Akiyama T et al (1999) PSD-95 promotes Fyn-mediates Tyrosine phosphorylation of the *N*-methyl-D-aspartate receptor subunit NR2A. Proc Natl Acad Sci U S A 96:435–440
156. Kalia LV, Salter MW (2003) Interactions between Src family protein tyrosine kinases and PSD-95. Neuropharmacol 45:720–728
157. Salter MW, Kalia LV (2004) Src kinases: a hub for NMDA receptor regulation. Nat Rev Neurosci 5:317–328
158. Nada S, Shima T, Yanai H et al (2003) Identification of PSD-93 as a substrate for the Src family tyrosine kinase Fyn. J Biol Chem 278:47610–47621
159. Jurd R, Tretter V, Walker J et al (2010) Fyn Kinase contributes to tyrosine phosphorylation of the GABA(A) receptor gamma2 subunit. Mol Cell Neurosci 44:129–134
160. Taniguchi S, Liu H, Nakazawa T et al (2003) p250GAP, A neural RhoGAP protein is associated with and phosphorylated by Fyn. Biochem Biophys Res Commun 306:151–155
161. Liang X, Draghi NA, Resh MD (2004) Signaling from integrins to Fyn to Rho family GTPase regulates morphologic differentiation of oligodendrocytes. J Neurosci 24:7140–7149
162. Lim SH, Kwon SK, Lee MK et al (2009) Synapse formation regulated by protein tyrosine phosphatase receptor T through interaction with cell adhesion molecules and Fyn. EMBO J 28:3564–3578
163. Wang PS, Wang J, Xiao ZC et al (2009) Protein-tyrosine phosphatase alpha acts as an upstream regulator of Fyn signaling to promote oligodendrocyte differentiation and myelination. J Biol Chem 284:33692–33702
164. LaFerla FM, Green KN, Oddo S (2007) Intracellular amyloid-beta in Alzheimer's disease. Nat Rev Neurosci 8:499–509
165. Haass C, Selkoe DJ (2007) Soluble protein oligomers in neurodegeneration: lessons from the Alzheimer's amyloid beta-peptide. Nat Rev Mol Cell Biol 8:101–112
166. Ballatore C, Lee VM, Trojanowski JQ (2007) Tau-mediated neurodegeneration in Alzheimer's disease and related disorders. Nat Rev Neurosci 8:663–672
167. Wang JZ, Liu F (2008) Microtubule-associated protein tau in development, degeneration and protection of neurons. Prog Neurobiol 85:148–175
168. Shirazi SK, Wood JG (1993) The protein tyrosine kinase, fyn, in Alzheimer's disease pathology. Neuroreport 4:435–437

169. Luo YQ, Hirashima N, Li YH et al (1995) Physiological levels of beta-amyloid increase tyrosine phosphorylation and cytosolic calcium. Brain Res 681:65–74
170. Klein WL, Krafft GA, Finch CE (2001) Targeting small Aβ oligomers: the solution to an Alzheimer's disease conundrum? Trends Neurosci 24:219–224
171. Chen S, Yadav SP, Surewicz WK (2010) Interaction between human prion protein and amyloid-beta (Aβ) oligomers: role of N-terminal residues. J Biol Chem 285:26377–26383
172. Calella AM, Farinelli M, Nuvolone M et al (2010) Prion protein and Aβ-related synaptic toxicity impairment. EMBO Mol Med 2:306–314
173. Balducci C, Beeg M, Stravalaci M et al (2010) Synthetic amyloid-beta oligomers impair long-term memory independently of cellular prion protein. Proc Natl Acad Sci U S A 107:2295–2300
174. Lauren J, Gimbel DA, Nygaard HB et al (2009) Cellular prion protein mediates impairment of synaptic plasticity by amyloid-beta oligomers. Nature 457:1128–1132
175. Um JW, Nygaard HB, Heiss JK et al (2012) Alzheimer amyloid-beta oligomer bound to postsynaptic prion protein activates Fyn to impair neurons. Nat Neurosci 15:1227–1235
176. Um JW, Kaufman AC, Kostylev M et al (2013) Metabotropic glutamate receptor 5 is a coreceptor for Alzheimer Aβ oligomer bound to cellular prion protein. Neuron 79:887–902
177. Roberson ED, Scearce-Levie K, Palop JJ et al (2007) Reducing endogenous tau ameliorates amyloid beta-induced deficits in an Alzheimer's disease mouse model. Science 316:750–754
178. Bhaskar K, Hobbs GA, Yen SH et al (2010) Tyrosine phosphorylation of tau accompanies disease progression in transgenic mouse models of tauopathy. Neuropathol Appl Neurobiol 36:462–477
179. Bhaskar K, Yen SH, Lee G (2005) Disease-related modifications in tau affect the interaction between Fyn and Tau. J Biol Chem 280:35119–35125
180. Lee G, Newman ST, Gard DL et al (1998) Tau interacts with src-family non-receptor tyrosine kinases. J Cell Sci 111:3167–3177
181. Larson M, Sherman MA, Amar F et al (2012) The complex PrP(c)–Fyn couples human oligomeric Aβ with pathological tau changes in Alzheimer's disease. J Neurosci 32:16857a-16871
182. Aarts M, Liu Y, Liu L et al (2002) Treatment of ischemic brain damage by perturbing NMDA receptor-PSD95 protein interactions. Science 298:846–850
183. Ittner LM, Ke YD, Delerue F et al (2010) Dendritic function of tau mediates amyloid-beta toxicity in Alzheimer's disease mouse models. Cell 298:846–850
184. Yun HM, Kim S, Kim HJ et al (2007) The novel cellular mechanism of human 5-HT6 receptor through an interaction with Fyn. J Biol Chem 282:5496–5505
185. Dubreuil P, Letard S, Ciufolini M (2009) Masitinib (AB1010), a potent and selective tyrosine kinase inhibitor targeting KIT. PLoS One 4, e7258
186. Marech I, Patruno R, Zizzo N et al (2014) Masitinib (AB1010), from canine tumor model to human clinical development: where we are? Crit Rev Oncol Hematol 91:98–111
187. Demetri GD (2011) Differential properties of current tyrosine kinase inhibitors in gastrointestinal stromal tumors. Semin Oncol 38:S10–S19
188. Mitry E, Hammel P, Deplanque G et al (2010) Safety and activity of masitinib in combination with gemcitabine in patients with advanced pancreatic cancer. Cancer Chemother Pharmacol 66:394–403
189. Niederhoffer N, Levy R, Sick E et al (2009) Amyloid beta peptides trigger CD47-dependent mast cell secretory and phagocytic responses. Int J Immunopathol Pharmacol 22:473–483
190. Maslinska D, Laure-Kamionowska M, Maslinski KT et al (2007) Distribution of tryptase-containing mast cells and metallothionein reactive astrocytes in human brains with amyloid deposits. Inflamm Res 5:S17–S18
191. Piette F, Belmin J, Vincent H et al (2011) Masitinib as an adjunct therapy for mild-to-moderate Alzheimer's disease: a randomised, placebo-controlled phase 2 trial. Alzheimers Res Ther 3:16

192. Hennequin LF, Allen J, Breed J et al (2006) N-(5-chloro-1,3-benzodioxol-4-yl)-7-[2-(4-methylpiperazin-1-yl)ethoxy]-5-(tetrahydro-2H-pyran-4-yloxy)quinazolin-4-amine, a novel, highly selective, orally available, dual-specific c-Src/Abl kinase inhibitor. J Med Chem 49:6465–6488
193. Gangadhar TC, Clark JI, Karrison T et al (2013) Phase II study of the Src kinase inhibitor saracatinib (AZD0530) in metastatic melanoma. Invest New Drugs 31:769–773
194. Gucalp A, Sparano JA, Caravelli J et al (2011) Phase II trial of saracatinib (AZD0530), an oral SRC-inhibitor for the treatment of patients with hormone receptor-negative metastatic breast cancer. Clin Breast Cancer 11:306–311
195. Mackay HJ, Au HJ, McWhirter E et al (2012) A phase II trial of the Src kinase inhibitor saracatinib (AZD0530) in patients with metastatic or locally advanced gastric or gastro esophageal junction (GEJ) adenocarcinoma: a trial of the PMH phase II consortium. Invest New Drugs 30:1158–1163
196. Fury MG, Baxi S, Shen R et al (2011) Phase II study of saracatinib (AZD0530) for patients with recurrent or metastatic head and neck squamous cell carcinoma (HNSCC). Anticancer Res 31:249–253
197. Renouf DJ, Moore MJ, Hedley D et al (2012) A phase I/II study of the Src inhibitor saracatinib (AZD0530) in combination with gemcitabine in advanced pancreatic cancer. Invest New Drugs 30:779–786
198. Hanke JH, Gardner JP, Dow RL et al (1996) Discovery of a novel, potent and Src family-selective tyrosine kinase inhibitor. Study of Lck- and Fyn T- dependent T cell activation. J Biol Chem 271:695–701
199. Schenone S, Zanoli S, Crullo C et al (2008) Current advances in the development of anticancer drugs targeting tyrosine kinases of the Src family. Curr Drug Ther 3:158–176
200. Tintori C, La Sala G, Vignaroli G et al (2015) Studies on the ATP binding site of Fyn kinase for the identification of new inhibitors and their evaluation as potential agents against tauopathies and tumors. J Med Chem 58:4590–4609
201. Ryan DA, Narrow WC, Federoff HJ et al (2010) An improved method for generating consistent soluble amyloid-beta oligomer preparations for in vitro neurotoxicity studies. J Neurosci Methods 190:171–179
202. Wong J, Higgins M, Halliday G et al (2012) Amyloid beta selectively modulates neuronal TrkB alternative transcript expression with implications for Alzheimer's disease. Neuroscience 210:363–374
203. Kinoshita T, Matsubara M, Ishiguro H et al (2006) Structure of human Fyn kinase domain complexed with staurosporine. Biochem Biophys Res Commun 346:840–844
204. Falchi F, Manetti F, Carraro F et al (2009) 3D QSAR models built on structure-based alignments of Abl tyrosine kinase inhibitors. ChemMedChem 6:976–987
205. Kruewel T, Schenone S, Radi M et al (2010) Molecular characterization of c-Abl/c-Src kinase inhibitors targeted against murine tumour progenitor cells that express stem cell markers. PLoS One 5, e14143
206. Radi M, Dreassi E, Brullo C et al (2011) Design, synthesis, biological activity, and ADME properties of pyrazolo[3,4-d]pyrimidines active in hypoxic human leukemia cells: a lead optimization study. J Med Chem 54:2610–2626
207. Schenone S, Bruno O, Bondavalli F et al (2004) Antiproliferative activity of new 1-aryl-4-amino-1H-pyrazolo[3,4-d]pyrimidine derivatives toward the human epidermoid carcinoma A431 cell line. Eur J Med Chem 39:939–946
208. Radi M, Brullo C, Crespan E et al (2011) Identification of potent c-Src inhibitors strongly affecting the proliferation of human neuroblastoma cells. Bioorg Med Chem Lett 21:5928–5933
209. Schindler T, Sicheri F, Pico A et al (1999) Crystal structure of Hck in complex with a Src family-selective tyrosine kinase inhibitor. Mol Cell 3:639–648

210. Zhu X, Kim JL, Newcomb JR et al (1999) Structural analysis of the lymphocyte-specific kinase Lck in complex with non-selective and Src family selective kinase inhibitors. Structure 7:651–661
211. Vignaroli G, Mencarelli M, Sementa D et al (2014) Exploring the chemical space around the privileged pyrazolo[3,4-d]pyrimidine scaffold: toward novel allosteric inhibitors of T315I-mutated Abl. ACS Comb Sci 16:168–175
212. Smith B, Medda F, Gokhale V et al (2012) Recent advances in the design, synthesis, and biological evaluation of selective DYRK1A inhibitors: a new avenue for a disease modifying treatment of Alzheimer's? ACS Chem Neurosci 3:857–872
213. Agholme L, Lindstrom T, Kagedal K et al (2010) An in vitro model for neuroscience: differentiation of SH-SY5Y cells into cells with morphological and biochemical characteristics of mature neurons. J Alzheimers Dis 20:1069–1082

Top Med Chem (2017) 24: 159–180
DOI: 10.1007/7355_2016_15

Published online: 11 December 2016

Microtubule-Stabilizing Agents for Alzheimer's and Other Tauopathies

Carlo Ballatore, Amos B. Smith III, Virginia M.-Y. Lee, John Q. Trojanowski, and Kurt R. Brunden

Abstract In neurodegenerative tauopathies, of which the most prevalent example is Alzheimer's disease (AD), the aggregation of the microtubule (MT)-stabilizing protein tau is believed to have neuropathological consequences. Multiple studies indicate that deficits in axonal MTs and axonal transport may contribute to the neurodegenerative processes of these diseases. MT-stabilizing molecules have shown promise in restoring axonal MTs and transport, as well as cognitive performance, in animal models of human tauopathies. As a result, such compounds may be considered as potential candidates for the treatment of AD and related tauopathies. Many examples of MT-stabilizing natural products and derivatives thereof have been approved for cancer treatment; however, the use of these compounds for central nervous system (CNS) diseases may be challenging due to limited brain penetration and oral bioavailability, as well as potential systemic side effects. In this chapter, we review the progress made toward the identification and development of CNS-active MT-stabilizing candidate compounds, with an emphasis on nonnaturally occurring small molecules that exhibit favorable drug-like properties.

C. Ballatore (✉)
Skaggs School of Pharmacy and Pharmaceutical Sciences, University of California, San Diego, 9500 Gilman Drive, MC0756, La Jolla, CA 92093, USA
e-mail: cballatore@ucsd.edu

A.B. Smith III
Department of Chemistry, School of Arts and Sciences, University of Pennsylvania, 231 South 34th Street, Philadelphia, PA 19104-6323, USA

V.M.-Y. Lee, J.Q. Trojanowski, and K.R. Brunden (✉)
Center for Neurodegenerative Disease Research, Institute on Aging, University of Pennsylvania, 3600 Spruce Street, Philadelphia, PA 19104-6323, USA
e-mail: kbrunden@upenn.edu.

Keywords Alzheimer's disease, Axonal transport, Epothilone D, Microtubule, Microtubule-stabilizing agent, Tau, Tauopathy, Triazolopyrimidines

Contents

1 Introduction

Neurodegenerative tauopathies [1] are a group of >20 sporadic and familial diseases (e.g., see Table 1) characterized by progressive dysfunction of nerve cells that correlates with the appearance of insoluble intracellular aggregates comprised primarily of tau proteins. Importantly, a causal relationship between the formation of tau inclusions and the process of neurodegeneration has been established via a series of genetic studies [2–4]. Among neurodegenerative tauopathies, the most prevalent example is Alzheimer's disease (AD), a complex multifactorial disease which features, in addition to tau inclusions, senile plaque pathology caused by the aggregation of amyloid β (Aβ) peptides.

Tau is a microtubule (MT)-associated protein that is highly expressed in the axons of neurons. The primary function of tau appears to be the promotion of MT assembly and stabilization [5]. As highlighted in Fig. 1, tau is comprised of: (1) either three or four alternatively spliced MT-binding amino acid repeats that interact with the β-tubulin subunit in a binding site that is believed to overlap the taxane-binding site [6]; (2) a proline-rich region, which at physiological pH is mostly positively charged and that is thought to interact electrostatically with the negatively charged surface of MTs; and (3) an N-terminal domain containing either zero, one, or two 29 amino acid-long alternatively spliced inserts, known as the projection domain region, which at physiological pH is mostly negatively charged and thus extends outward, away from the surface of the MT due to electrostatic repulsion. In the adult brain, there are six tau isoforms that result from the aforementioned alternative splicing of the MT-binding repeats (i.e., 3R or 4R) and the N-terminal inserts (i.e., 0N, 1N, or 2N). In the healthy adult brain, the ratio of 3R and 4R tau is close to 1:1, and a deviation from this ratio is associated with certain forms of frontotemporal lobar degeneration (FTLD).

Tau is known to undergo a number of posttranslational modifications [7] that are believed to play an important role in modulating binding to MTs, as well as the propensity of the protein to undergo misfolding and aggregation. In neurodegenerative diseases, tau is typically hyperphosphorylated and aggregated in neurons, forming either paired helical filaments (PHF) or straight filaments that comprise

Table 1 Selected examples of neurodegenerative amyloidoses featuring tau inclusions

Alzheimer's disease
Chronic traumatic encephalopathy
Corticobasal degeneration
Frontotemporal dementia and parkinsonism linked to chromosome 17
Frontotemporal lobar degeneration
Pick's disease
Progressive supranuclear palsy

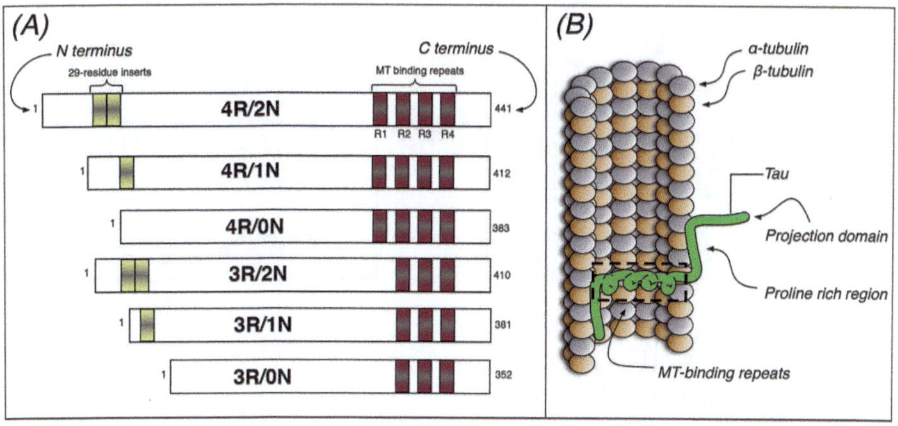

Fig. 1 (**a**) The six isoforms of tau; (**b**) A schematic model of the tau MT complex revealing key interactions (adapted from Amos [6])

neurofibrillary tangles (NFTs) in the cell soma and neuritic threads in neuronal processes [1, 5]. The degree of cognitive impairment in AD appears to correlate closely with the density of NFTs, although a small number NFTs can also form as part of normal aging [8–10]. Furthermore, in the diseased state, aggregated tau is comparatively hyperphosphorylated relative to aggregated tau in non-diseased adult brains [11].

A growing body of evidence indicates that a common feature of neurodegenerative tauopathies is compromised axonal transport that appears to be associated with a faulty axonal MT network [12]. Axonal transport is an essential process in neurons that is responsible for the movement of various types of cargoes to and from the cell body and the synapses [13]. MTs are essential constituents of the intracellular infrastructure, and they form polarized linear arrays in the axons of neurons that provide both structural support and directionality to motor proteins, such as kinesins and dyneins (see Fig. 2). Thus, an appropriate structure and organization of axonal MTs is believed to be required for efficient axonal transport. In addition, there is evidence that alterations in MT dynamics in other neuronal compartments may play a role in the neurodegenerative process. For example,

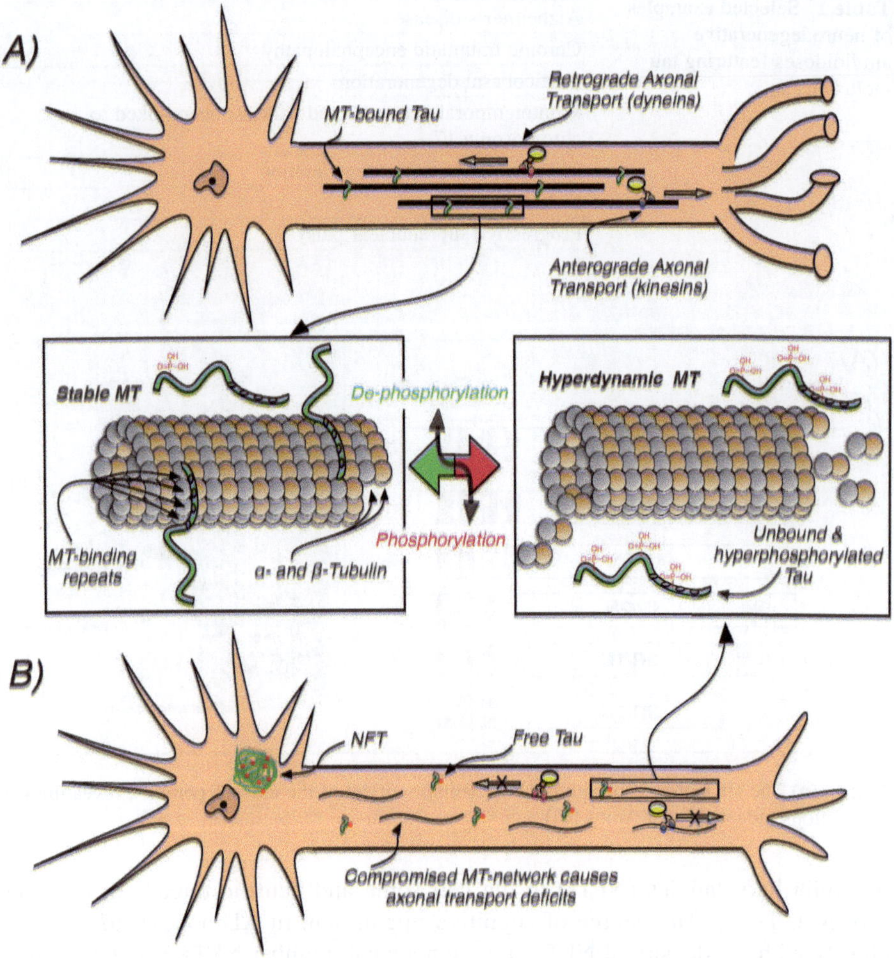

Fig. 2 Schematic representation of the role of tau, axonal MTs, and axonal transport under physiological conditions (**a**). The hyperphosphorylation and disengagement of tau from MTs result in compromised axonal transport, leading to a pathological condition (**b**)

abnormal MT dynamics in the dendrites may be involved in the loss of dendritic spines that constitute another distinctive neuropathological feature of AD [14–17]. Interestingly, a decreased level of biomarkers of stable MTs, such as acetylated and de-tyrosinated α-tubulin [18, 19], as well as a reduction in the total number and lengths of neuronal MTs, have been observed in the brains of AD patients [20–22], suggesting that hyperdynamic MTs, axonal transport deficits, and neurodegeneration may be causally related.

Mechanistically, the alteration of axonal MTs in tauopathy neurons may derive at least in part from the fact that tau, when hyperphosphorylated and sequestered into aggregate structures, can no longer bind to and stabilize MTs [23]. Evidence of a causative relationship between the hyperphosphorylated/aggregated tau and disrupted MTs has been found in cell culture [24, 25], as well as in animal models [26–28]. The binding affinity of tau for MTs is influenced by the phosphorylation state of the protein [29], as increased phosphorylation is believed to reduce the binding affinity of tau for MTs [30]. Under normal conditions, tau exhibits a relatively low phosphorylation status (i.e., an average of ~1.9 mol of phosphate per mole of tau [31]) and is mostly bound to MTs [32]. In the AD brain, however, the tau phosphorylation increases to ~2.6 mol of phosphate per mole of tau, with PHF tau exhibiting 6–8 mol of phosphate per mole of tau [31]. Thus, the hyperphosphorylation and aggregation of tau are likely to cause an abnormal disengagement of tau from MTs that may lead to cytoskeletal alterations [33] and axonal transport deficits in affected neurons. In addition, as there is evidence that MT-bound tau may also be involved in regulating the attachment of motor proteins, an aberrant detachment of tau from MTs could also affect axonal transport by interfering with normal motor protein function [34]. Finally, it is possible that tau hyperphosphorylation and aggregation may contribute to neuronal dysfunction by additional mechanisms that are not directly related to a loss of MT-stabilizing function but rather are due to a gain-of-function toxicity of misfolded tau monomers, oligomers, or fibrils.

2 MT-Stabilizing Agents as Candidate Therapeutics

Naturally Occurring Compounds and Derivatives Thereof Stabilization of axonal MTs by means of MT-stabilizing agents has been proposed as a possible strategy for therapeutic intervention in neurodegenerative tauopathies [35]. A series of proof-of-concept studies [26–28, 36] in animal models have provided validation for this strategy. The first important demonstration that MT-stabilizing drugs could be used to ameliorate MT and axonal transport deficits was obtained in 2005 [28], where weekly doses of a formulated paclitaxel (Taxol®, **1**, Fig. 3) preparation were found to restore axonal transport and reduce the neurodegenerative phenotype of T44 tau transgenic mice.

Although paclitaxel is known to have limited brain uptake, efficacy was presumably obtained in the T44 mice because they exhibit tau pathology in motor neurons that reside in the periphery. As a result, blood–brain barrier (BBB) penetration was not a requirement in these particular studies. Subsequent studies from our group [27, 36] and others [26] that utilized the brain-penetrant natural product, epothilone D (**2**, Fig. 4), in transgenic mouse models in which tau pathology is confined predominantly to the brain further validated the potential of MT-stabilizing agents for the treatment of tauopathies.

Several reports, including competition binding [37], X-ray [38], NMR [39], and photoaffinity-labeling [40] studies demonstrated that the epothilones, like other classes of MT-stabilizing natural products, interact with MTs at the taxane-binding

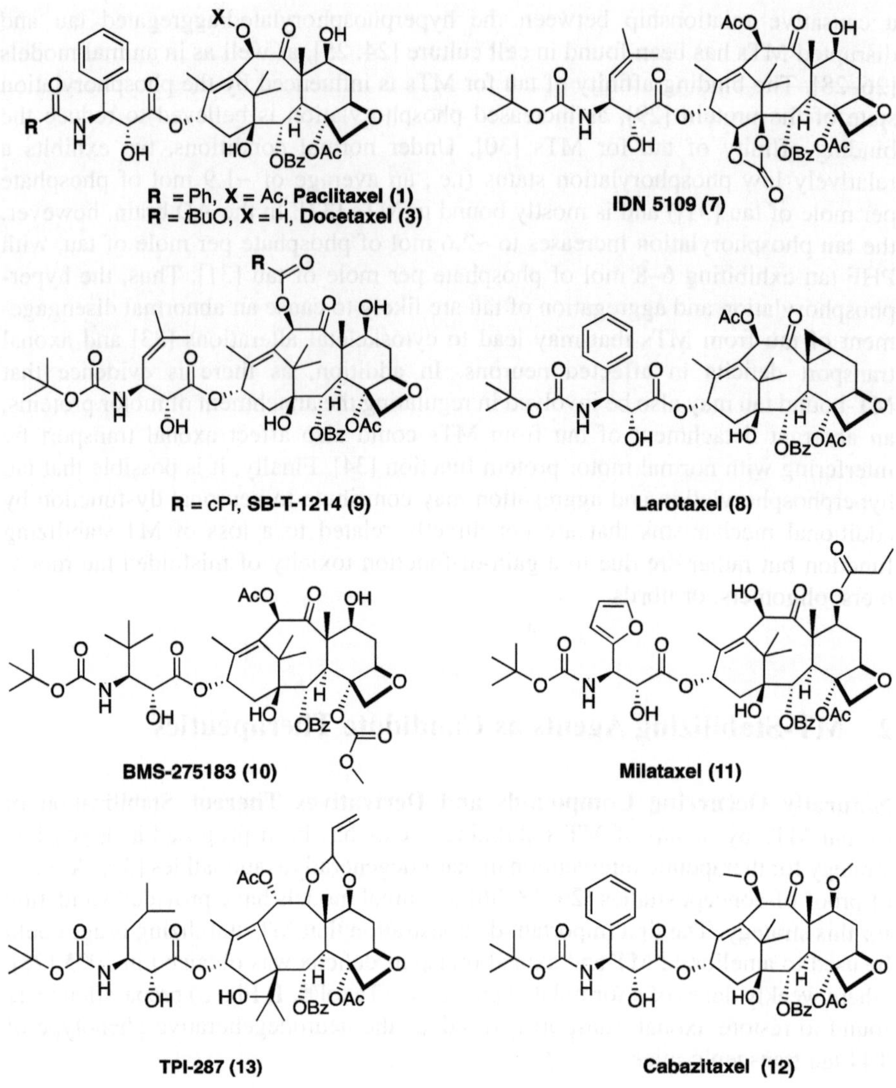

Fig. 3 The structure of paclitaxel and related taxanes

site located in the lumen of the MT in the β-tubulin subunit. The different taxane site MT-stabilizing agents are believed to promote MT stabilization by inducing a similar structuring of the M-loop of β-tubulin [38], which results in stronger lateral interactions between adjacent protofilaments. However, at the molecular level, the mechanism of action of these compounds is not identical due to differing binding modes. Indeed, even relatively small structural differences within the same class of compounds, such as those between paclitaxel and docetaxel (**3**, Fig. 3), are known to be responsible for noticeable changes in the structure of MTs [41].

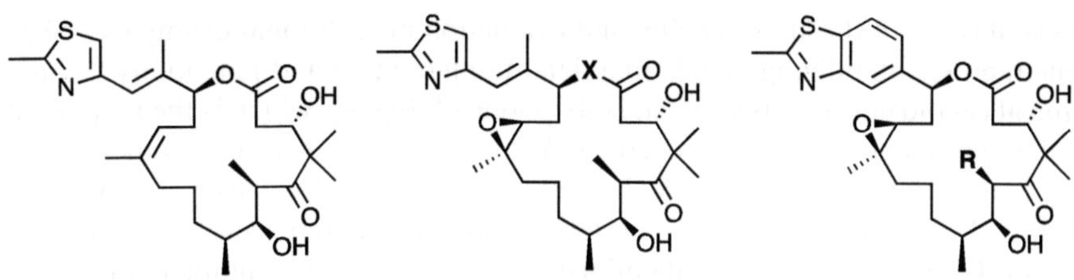

Fig. 4 The structures of selected epothilones

Paclitaxel exhibits limited brain penetration in part because of the active transporter, P-glycoprotein (P-gp) [42], which is highly expressed in the endothelial cells of the BBB. Epothilone D, unlike **1**, is not a good substrate for P-gp, but it is rather a potent inhibitor of the active transporter [43]. The ability of epothilone D to overcome P-gp-mediated efflux, combined with the fact that this molecule is comparatively smaller and less functionalized than paclitaxel and related taxanes, provides a possible explanation for the superior brain penetration of this natural product [43]. Treatment of tau transgenic mice, such as the PS19, 3xTg, and rTg4510 models that progressively develop tau pathology in the CNS with age, with relatively low weekly doses of epothilone D led to a reduction in the number of dystrophic axons and a reduction in tau pathology that coincided with a normalization of MT mass and dynamics, as well as an improvement in axonal transport [26, 27, 36]. Importantly, epothilone D treatment led to an improvement in cognitive performance and a reduction in neuronal loss in the tau transgenic mice. Interestingly, the effective doses of epothilone D in these studies (i.e., 0.3 and 1 mg/kg) were approximately equivalent to 1/30th and 1/100th of the chemotherapeutic doses when adjusted by allometric scaling for species differences. Treatment of tau transgenic animals with these relatively low weekly doses of epothilone D did not result in detectable side effects, such as neutropenia or peripheral neuropathy [44] that are typically encountered with chemotherapeutic regimens based on MT-stabilizing drugs.

Given the very promising results from these preclinical studies, the development of CNS-directed MT-stabilizing therapies holds promise as a possible strategy to treat tau-mediated neurodegeneration. Although epothilone D may be considered as a potentially promising clinical candidate, as this compound progressed to Phase 1b clinical testing in a short 9-week study with AD patients, **2** is only one of very few examples of brain-penetrant MT-stabilizing agents to be evaluated in animal models. Moreover, this compound may have potential deficiencies as a drug candidate to be used for a prolonged treatment period, including an IV route of administration and inhibition of P-gp, which may lead to undesired drug-drug interactions [45] and/or to CNS toxicities [46].

Over the past several years, a growing number of MT-stabilizing compounds have been identified among natural products and nonnaturally occurring small molecules [47, 48]. The screening and evaluation of additional examples of brain-penetrant MT-stabilizing agents may lead to alternative and potentially improved clinical candidates. In this context, ixabepilone (**4**, Fig. 4), which is the first member of the epothilones to have reached FDA approval for cancer chemotherapy [49], was evaluated in our laboratories. Ixabepilone is a close derivative of epothilone B (**5**, Fig. 4), in which the lactone moiety of the latter is replaced by a more stable lactam. Like epothilone D, ixabepilone was found to reach significant brain concentrations in mice after peripheral administration [43]. However, in stark contrast with the former compound, treatment of PS19 tau transgenic mice with **4** over several weeks was associated with adverse side effects at doses required to elicit an increase in a marker of stable MTs in the brain (data not published), suggesting that the repurposing of this drug for neurodegenerative diseases may not be possible. Interestingly, although both **2** and **4** exhibit prolonged half-lives in the brain (i.e., >16 h), epothilone D was found to be considerably less stable toward metabolism than ixabepilone, as suggested by a relatively short plasma half-life in mice (i.e., <2 h) [43]. A possible explanation for the noticeable difference in tolerability between these two compounds may be related in part to the fact that epothilone D is more rapidly metabolized outside of the brain and, therefore, has a lesser peripheral exposure than ixabepilone. This possibility may imply that a differential between brain and plasma half-life could be an important aspect to consider in the prioritization of candidate compounds for tauopathy treatment. A similar differential has been observed for other members of the epothilones [43], including sagopilone [50] (**6**, Fig. 4) and epothilone B [51]. Interestingly, epothilone B showed promise as a CNS-active agent in animal models of Parkinson's disease [52] and traumatic brain injury [53], with no signs of systemic toxicity.

Among the taxanes, different examples of paclitaxel derivatives have been identified that unlike **1** are capable of overcoming P-gp-mediated efflux and as a result may exhibit improved brain penetration and/or oral bioavailability (see compounds **7–12**, Fig. 3). These derivatives include IDN-5109 (**7**) [54], lorataxel (**8**) [55], SB-T-1214 (**9**) [56], BMS-275183 (**10**) [57, 58], and milataxel (**11**) [59]. Cabazitaxel (Jevtana®, **12**, Fig. 3) [60, 61], an FDA-approved anticancer drug, was also reported to exhibit increased brain penetration due to reduced transport by P-gp [62]; however, recent pharmacokinetic studies in P-gp-expressing and knockout mice clearly suggest that the brain accumulation of **12** may still be limited by the action of P-gp [63]. Another compound that was found to be particularly promising as a CNS-directed MT-stabilizing compound is TPI-287 (**13**, Fig. 3) [64]. This compound is a taxoid derivative that features a 11(15→1) abeotaxane scaffold, which is found also in several other naturally occurring taxoids from *Taxus chinensis*, *T. yunnanensis*, and *T. wallichiana* [65, 66]. This particular abeotaxane skeleton is characterized by a five-membered A ring of the baccatin core, compared to the larger six-membered ring found in all other taxanes. As TPI-287 was found to be brain penetrant, this compound is under evaluation as potential treatment for brain cancer and primary tumors that have metastasized to the brain [67–69]. In addition, although the activity of this compound in tau transgenic animal models has not been fully described, preliminary studies suggest that low weekly doses of TPI-287 may improve behavioral impairment and tau pathology in PS19 mice [70]. Importantly, this compound is presently undergoing clinical trials in patients with tauopathies, such as corticobasal degeneration and progressive supranuclear palsy (see clinicaltrials.gov identifier NCT02133846).

In addition to taxanes and epothilones, other classes of MT-stabilizing natural products are being considered for evaluation. For example, dictyostatin [71] (**14**, Fig. 5), another compound that is known to interact with MTs at the taxane site within β-tubulin [72] and which demonstrated promising pharmacokinetic and pharmacodynamic properties that resembled those of epothilone D [73], has recently undergone evaluation in PS19 tau transgenic mice in our laboratories [74]. Moreover, peloruside A [75] (**15**, Fig. 5) has also been described as a potentially promising compound, as in vitro studies demonstrated that this MT-stabilizing agent is able to compensate for the loss of MT-stabilizing function of tau caused by okadaic acid-induced hyperphosphorylation of tau [76]. These results with peloruside A are especially interesting in consideration of the fact that this compound stabilizes MTs in a substantially different way than molecules that bind the taxane site on β-tubulin, interacting instead with a distinct site on the surface of MTs that does not overlap with the taxane site [77, 78]. The brain pharmacokinetic properties of this compound, however, have not yet been determined.

Nonnaturally Occurring Small Molecules Most of the MT-stabilizing compounds discovered to date comprise a large ensemble of structurally complex natural products and derivatives thereof [47, 48]. With the possible exception of the selected examples illustrated above, the majority of these compounds share similar potential limitations as therapeutic candidates for CNS indications, including suboptimal brain exposure and, in some instances, synthetic challenges. For these reasons, the development of MT-stabilizing small molecules that exhibit more favorable drug-like properties would be desirable to identify candidates that could combine synthetic accessibility with good pharmaceutical properties, including oral bioavailability and brain penetration. It is known that nonnaturally occurring small molecules having MT-stabilizing activity are relatively rare, especially compared

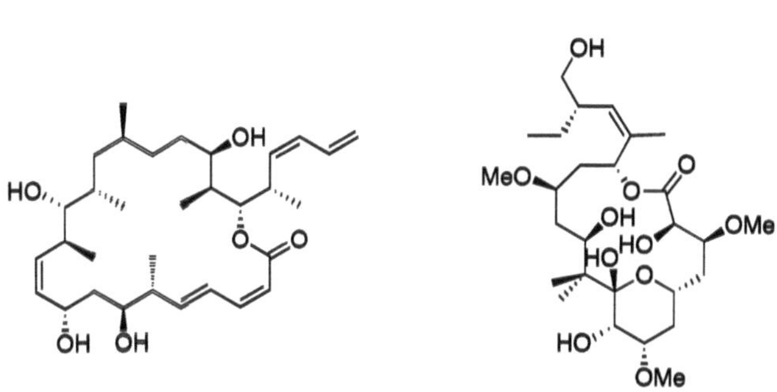

Dictyostatin (14) **Peloruside A (15)**

Fig. 5 Other MT-stabilizing natural products being investigated in the context of tau-mediated neurodegeneration

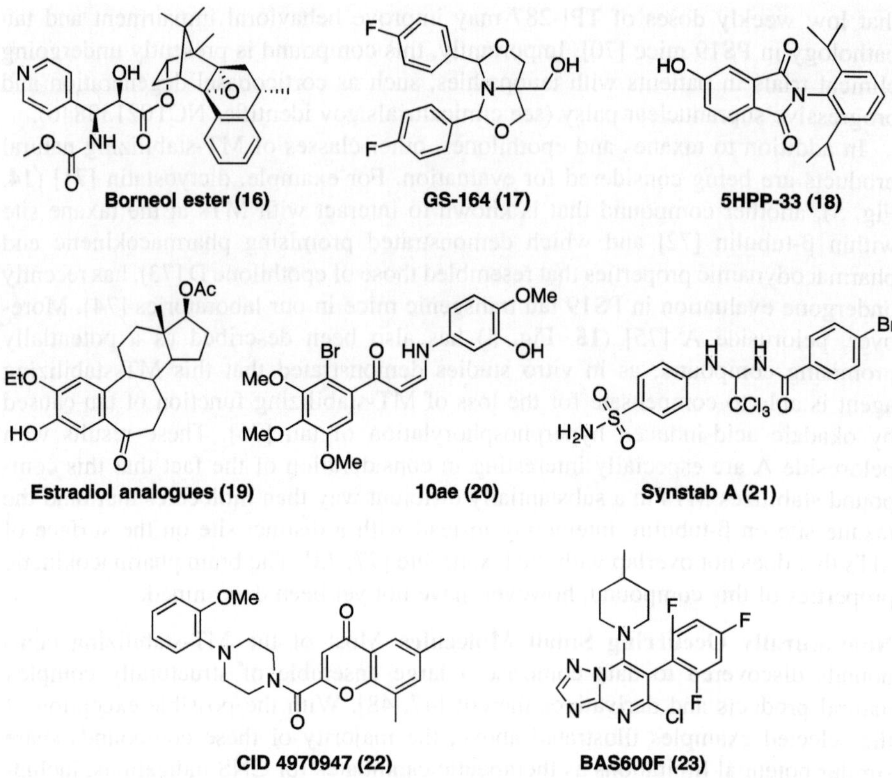

Fig. 6 Representative structures of nonnaturally occurring MT-stabilizing compounds

to compounds having the opposite effect of de-polymerizing MTs [79]. An overview of different classes of nonnaturally occurring compounds with reported MT-stabilizing activity is shown in Fig. 6 and Table 2.

Attempts to simplify the taxane structure and generate smaller and less complex analogs of **1** with paclitaxel-like MT-stabilizing properties resulted in the identification of borneol ester derivatives [94] (e.g., **16**, Fig. 6) and GS-164 [95] (**17**, Fig. 6). Interestingly, although these compounds have been found to promote MT polymerization in vitro in a manner similar to paclitaxel, compared to the natural product, these structurally simpler analogs appear to be considerably less potent as antimitotic agents. This observation is somewhat paradoxical, as it is thought that the antimitotic activity of MT-stabilizing agents directly results from alterations of MTs in the mitotic spindle during cell division. If the apparent separation between MT-stabilizing and antimitotic activities is further substantiated for these compounds, they may serve as potentially attractive starting points for the treatment of neurodegenerative diseases [94].

Relatively weak antimitotic and MT-stabilizing properties have also been described for a group of thalidomide derivatives [96]. The most studied of these

Table 2 Representative examples of triazolopyrimidines and related heterocyclic compounds

Compound class/name	Representative structure	References
Triazolopyrimidines	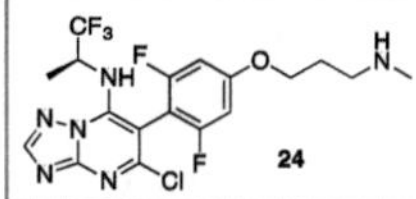 24	[80–83]
	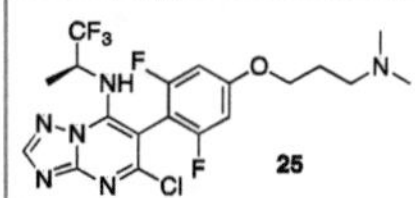 25	[83–85]
	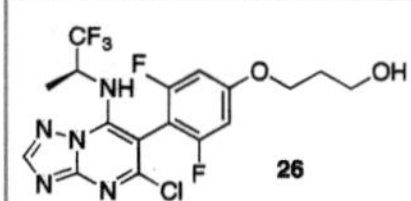 26	[83–86]
	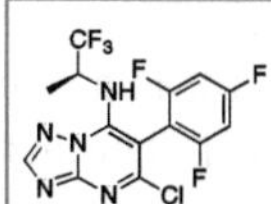 27	[83–85]
	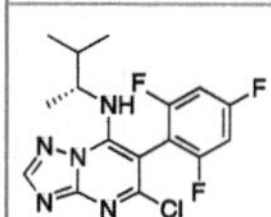 28	[84]
Phenylpyrimidines	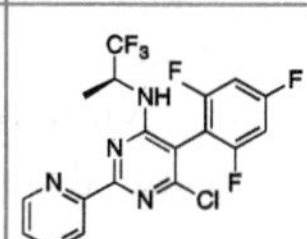 29	[84, 87, 88]
Pyridazines	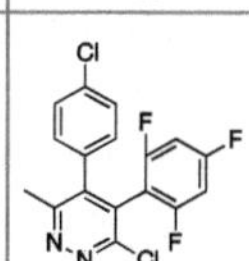 30	[89]
Pyridopyrazines	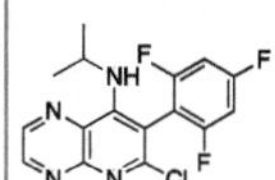 31	[90]
Imidazoles	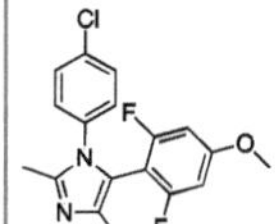 31	[91]

(continued)

Table 2 (continued)

Compound class/name	Representative structure	References
Pyrazoles	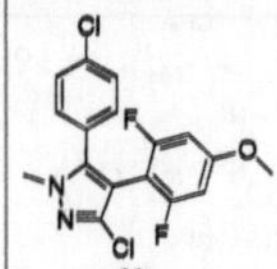 32	[92]
Pyrazinones	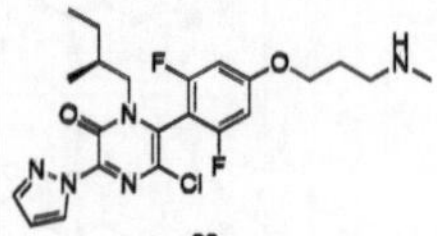 33	[93]

derivatives, 5HPP-33 [97] (**18**, Fig. 6), is a compound that was initially identified for its anti-angiogenic properties [98]. The activity of **18**, however, is somewhat unclear as previous studies found that this compound exhibits inhibitory activity on tubulin polymerization [99]. In addition, recent studies [100] revealed that although 5HPP-33 suppresses the dynamics of individual MTs in MCF-7 cells, this compound causes an inhibition of MT assembly both in cell-free assays and in cultured MCF-7 cells.

Selected estradiol derivatives have also been identified that appear to promote MT polymerization [101]. Interestingly, the metabolite of estradiol, 2-methoxyestradiol, and other related synthetic congeners are known to act as inhibitors of MT polymerization and are thought to interact with tubulin at the binding site for colchicine [102], an inhibitor of MT polymerization. However, selected congeners (e.g., **19**, Fig. 6) demonstrated paclitaxel-like enhancement of MT polymerization [103]. Structure-activity relationship (SAR) studies demonstrated that the presence in these compounds of an acetyl group at C17, but not at C3, is essential to impart MT-stabilizing properties [101, 103].

In an apparently similar fashion to the estradiol derivatives, a series of enaminones, such as 10ae (**20**, Fig. 6), have also been identified that appear to be functionally related to paclitaxel with respect to MT stabilization and antimitotic activity [104]. However, several structurally related compounds have been found to compete with colchicine binding to tubulin and inhibit tubulin polymerization [105]. The tubulin polymerization promoter, 10ae, was found to be more potent than paclitaxel in a cell-free, light-scattering MT polymerization assay; however, in cells, this compound was not as potent. Furthermore, significant differences in the organization of cytoplasmic MTs have been noted upon treatment with paclitaxel or **20**, as cells treated with the latter compound did not show the characteristic bundling of MTs that is typically seen upon treatment with taxanes [104]. These results with enaminone derivatives and those mentioned above with estradiol derivatives demonstrate that appropriate structural modifications of certain colchicine-site tubulin polymerization inhibitors can lead to analogs with opposite

MT-stabilizing activities. It is not clear, however, whether these MT-stabilizing analogs interact with tubulin at the colchicine site or at a different binding site.

Additional compounds of interest have been identified via phenotypic screening for novel small molecules having paclitaxel-like properties. These compounds include synstab A [79] (**21**, Fig. 6) and CID 4970947 [106] (**22**, Fig. 6), which demonstrated MT-stabilizing activity both in cell-based and cell-free assays. The SAR of these compounds, however, has not yet been investigated. Finally, a promising class of tubulin polymerization promoters are the triazolopyrimidines (e.g., BAS600F [107], **23**, Fig. 6) and related heterocycles (Table 2). This class of small heterocyclic compounds was originally identified via screening for novel antifungal agents. Considerable efforts, led largely by the agrochemical industry, resulted in the identification of a growing number of mono- and bi-heterocyclic compounds that are structurally related to the triazolopyrimidines and which exhibit comparable activities as tubulin polymerization promoters and antifungal agents (e.g., see compounds **25–34**, Table 2). Interestingly, Wyeth developed a series of triazolopyrimidine congeners as potential anticancer agents, and this effort ultimately led to the identification of TTI-237 (cevipabulin, **24**, Table 2) as a preferred clinical candidate with a promising combination of antimitotic activity and drug-like properties, including metabolic stability and oral bioavailability [83]. Importantly, although cevipabulin was found to exhibit limited brain uptake, several examples of MT-stabilizing congeners (e.g., **25–29**, Table 2) have been identified that are brain penetrant and that can cause an elevation in biomarkers of stable MTs in the brain of wild-type (WT) mice [84, 85].

Early studies to elucidate the mode of action and the SAR of triazolopyrimidines indicated that these compounds alter MT dynamics in a manner potentially unique among all MT-active compounds [82, 83]. Competition binding studies revealed that cevipabulin does not compete with colchicine or paclitaxel for binding to tubulin. Instead, **24** appeared to inhibit vinblastine binding, suggesting that the triazolopyrimidine class of compounds may be competing directly or allosterically with the binding site of vinca alkaloids, which inhibit MT polymerization. This observation, combined with the fact that cells treated with cevipabulin manifested changes in the level of different phosphoproteins that are typically seen after vincristine treatment, led to the hypothesis that the mode of action of triazolopyrimidines may be both taxane- and vinca-like [82].

Further in vitro characterization of the pharmacological properties of cevipabulin, as well as the brain-penetrant derivatives **25** and **26** and phenylpyrimidine **29**, revealed that compound concentrations that are needed to elicit an increase in markers of stable MTs (i.e., acetylated- and de-tyrosinated-α-tubulin) in cells (QBI-293 or COS-7 cells) also produce a decrease in total α- and β-tubulin levels [84]. Notably, this compound-induced reduction of total tubulin, which occurs via the proteasomal degradation pathway, was not observed upon treatment with epothilone D or vinblastine. Furthermore, side-by-side comparison of cellular response after treatment with epothilone D, **25**, or **29** demonstrated that while all compounds produced a marked increase in acetylated α-tubulin, treatment with **25** or **29**, unlike epothilone D, resulted in altered distribution and intensity of β-tubulin labeling.

These results clearly suggest that while triazolopyrimidine **25** and phenylpyrimidine **29** can promote tubulin assembly, as determined by an increase in acetylated α-tubulin, the MT structures generated in the presence of these compounds may be abnormal and subject to degradation, an effect that would be undesirable for tauopathy treatment. However, an evaluation of a series of closely related congeners, such as triazolopyrimidine **27** and **28**, revealed that these compounds promote MT assembly without affecting total tubulin levels or disrupting MT architecture. These studies highlight that, depending on the particular scaffold and the nature of the substituents decorating the heterocycle, compound treatment can elicit substantially different cellular phenotypes [84]. Importantly, most of the preferred subset of MT-stabilizing triazolopyrimidines were found to be both brain penetrant and orally bioavailable and did not disrupt P-gp function. A representative example (**28**, Table 2) was also found to produce an elevation of markers of stable MTs in the brain of WT mice. Moreover, in neuronal culture, **28** could compensate for the loss of tau MT-stabilizing function caused by okadaic acid-induced tau hyperphosphorylation [84]. Thus, collectively, these results suggest that selected triazolopyrimidines and possibly other related heterocycles may provide a source of viable candidates for tauopathy treatment.

3 Concluding Remarks

CNS-active MT-stabilizing compounds hold promise as potential treatments for a variety of devastating conditions, including brain tumors and several neurodegenerative diseases [108]. Among natural products and related derivatives, selected compounds, such as epothilone D and TPI-287, have reached the early stages of clinical evaluation for neurodegenerative tauopathies. In addition, in spite of the fact that, historically, the identification of drug-like small molecules endowed with MT-stabilizing activity has proven to be challenging, over the past several years, there has been considerable progress as evidenced by the number of compounds that have been reported, most of which are related to the triazolopyrimidine class. Although the mode of action of these nonnaturally occurring compounds has not yet been completely defined, the data generated thus far clearly suggest that a select subset of these molecules may be viable candidates for neurodegenerative tauopathies and perhaps CNS oncological indications. Further evaluation of efficacy and tolerability in disease models will help determining the potential of these compounds as therapeutic agents.

Acknowledgments These studies were supported by NIH grants AG17586, AG029213, and AG044332. We would also like to acknowledge the generous donations made by the Karen Cohen Segal and Christopher S. Segal Alzheimer Drug Discovery Initiative Fund, the Paula C. Schmerler Fund for Alzheimer's Research, the Barrist Neurodegenerative Disease Research Fund, the Eleanor Margaret Kurtz Endowed Fund, the Mary Rasmus Endowed Fund for Alzheimer's Research, and Mrs. Gloria J. Miller and Arthur Peck, M.D., the Wood Foundation.

References

1. Lee VMY, Goedert M, Trojanowski JQ (2001) Neurodegenerative tauopathies. Annu Rev Neurosci 24(1):1121–1159
2. Hutton M, Lendon CL, Rizzu P, Baker M, Froelich S, Houlden H, Pickering-Brown S, Chakraverty S, Isaacs A, Grover A, Hackett J, Adamson J, Lincoln S, Dickson D, Davies P, Petersen RC, Stevens M, de Graaff E, Wauters E, van Baren J, Hillebrand M, Joosse M, Kwon JM, Nowotny P, Che LK, Norton J, Morris JC, Reed LA, Trojanowski J, Basun H, Lannfelt L, Neystat M, Fahn S, Dark F, Tannenberg T, Dodd PR, Hayward N, Kwok JB, Schofield PR, Andreadis A, Snowden J, Craufurd D, Neary D, Owen F, Oostra BA, Hardy J, Goate A, van Swieten J, Mann D, Lynch T, Heutink P (1998) Association of missense and 5'-splice-site mutations in tau with the inherited dementia FTDP-17. Nature 393(6686):702–705. doi:10.1038/31508
3. Poorkaj P, Bird TD, Wijsman E, Nemens E, Garruto RM, Anderson L, Andreadis A, Wiederholt WC, Raskind M, Schellenberg GD (1998) Tau is a candidate gene for chromosome 17 frontotemporal dementia. Ann Neurol 43(6):815–825. doi:10.1002/ana.410430617
4. Spillantini MG, Murrell JR, Goedert M, Farlow MR, Klug A, Ghetti B (1998) Mutation in the tau gene in familial multiple system tauopathy with presenile dementia. Proc Natl Acad Sci U S A 95(13):7737–7741
5. Buee L, Bussiere T, Buee-Scherrer V, Delacourte A, Hof PR (2000) Tau protein isoforms, phosphorylation and role in neurodegenerative disorders. Brain Res Rev 33(1):95–130
6. Amos LA (2004) Microtubule structure and its stabilisation. Org Biomol Chem 2(15): 2153–2160
7. Gong C-X, Liu F, Grundke-Iqbal I, Iqbal K (2004) Post-translational modifications of tau protein in Alzheimer's disease. J Neural Transm 112(6):813–838. doi:10.1007/s00702-004-0221-0
8. Arriagada PV, Growdon JH, Hedley-Whyte ET, Hyman BT (1992) Neurofibrillary tangles but not senile plaques parallel duration and severity of Alzheimer's disease. Neurology 42(3): 631–639
9. Arriagada PV, Marzloff K, Hyman BT (1992) Distribution of Alzheimer-type pathologic changes in nondemented elderly individuals matches the pattern in Alzheimer's disease. Neurology 42(9):1681–1688
10. Guillozet AL, Weintraub S, Mash DC, Mesulam M (2003) Neurofibrillary tangles, amyloid, and memory in aging and mild cognitive impairment. Arch Neurol 60(5):729–736. doi:10.1001/archneur.60.5.729
11. D'Souza I, Schellenberg GD (2005) Regulation of tau isoform expression and dementia. Biochim Biophys Acta (BBA) - Mol Basis Dis 1739(2–3):104–115. doi:10.1016/j.bbadis.2004.08.009
12. Millecamps S, Julien J-P (2013) Axonal transport deficits and neurodegenerative diseases. Nat Rev Neurosci 14(3):161–176
13. Hirokawa N, Takemura R (2005) Molecular motors and mechanisms of directional transport in neurons. Nat Rev Neurosci 6(3):201–214
14. Gu J, Firestein BL, Zheng JQ (2008) Microtubules in dendritic spine development. J Neurosci 28(46):12120–12124. doi:10.1523/jneurosci.2509-08.2008
15. Hoogenraad CC, Akhmanova A (2010) Dendritic spine plasticity: new regulatory roles of dynamic microtubules. Neuroscientist 16(6):650–661. doi:10.1177/1073858410386357

16. Jaworski J, Hoogenraad CC, Akhmanova A (2008) Microtubule plus-end tracking proteins in differentiated mammalian cells. Int J Biochem Cell Biol 40(4):619–637. doi:10.1016/j.biocel.2007.10.015
17. Penazzi L, Tackenberg C, Ghori A, Golovyashkina N, Niewidok B, Selle K, Ballatore C, Smith AB III, Bakota L, Brandt R (2016) Abeta-mediated spine changes in the hippocampus are microtubule-dependent and can be reversed by a subnanomolar concentration of the microtubule-stabilizing agent epothilone D. Neuropharmacology 105:84–95. doi:10.1016/j.neuropharm.2016.01.002
18. Black MM, Baas PW, Humphries S (1989) Dynamics of alpha-tubulin deacetylation in intact neurons. J Neurosci 9(1):358–368
19. Laferriere NB, MacRae TH, Brown DL (1997) Tubulin synthesis and assembly in differentiating neurons. Biochem Cell Biol 75(2):103–117
20. Cash AD, Aliev G, Siedlak SL, Nunomura A, Fujioka H, Zhu X (2003) Microtubule reduction in Alzheimer's disease and aging is independent of tau filament formation. Am J Pathol 162(5):1623–1627
21. Hempen B, Brion J-P (1996) Reduction of acetylated α-tubulin immunoreactivity in neurofibrillary tangle-bearing neurons in Alzheimer's disease. J Neuropathol Exp Neurol 55(9):964–972. doi:10.1097/00005072-199609000-00003
22. Zhang F, Su B, Wang C, Siedlak SL, Mondragon-Rodriguez S, Lee H-g, Wang X, Perry G, Zhu X (2015) Posttranslational modifications of α-tubulin in alzheimer disease. Transl Neurodegener 4(1):1–9. doi:10.1186/s40035-015-0030-4
23. Ballatore C, Lee VMY, Trojanowski JQ (2007) Tau-mediated neurodegeneration in Alzheimer's disease and related disorders. Nat Rev Neurosci 8(9):663–672
24. Alonso AC, Zaidi T, Grundke-Iqbal I, Iqbal K (1994) Role of abnormally phosphorylated tau in the breakdown of microtubules in Alzheimer disease. Proc Natl Acad Sci U S A 91(12):5562–5566
25. Merrick SE, Trojanowski JQ, Lee VMY (1997) Selective destruction of stable microtubules and axons by inhibitors of protein serine/threonine phosphatases in cultured human neurons. J Neurosci 17(15):5726–5737
26. Barten DM, Fanara P, Andorfer C, Hoque N, Wong PYA, Husted KH, Cadelina GW, DeCarr LB, Yang L, Liu L, Fessler C, Protassio J, Riff T, Turner H, Janus CG, Sankaranarayanan S, Polson C, Meredith JE, Gray G, Hanna A, Olson RE, Kim SH, Vite GD, Lee FY, Albright CF (2012) Hyperdynamic microtubules, cognitive deficits, and pathology are improved in tau transgenic mice with low doses of the microtubule-stabilizing agent BMS-241027. J Neurosci 32(21):7137–7145
27. Zhang B, Carroll J, Trojanowski JQ, Yao Y, Iba M, Potuzak JS, Hogan AM, Xie SX, Ballatore C, Smith AB III, Lee VM, Brunden KR (2012) The microtubule-stabilizing agent, epothilone D, reduces axonal dysfunction, neurotoxicity, cognitive deficits, and Alzheimer-like pathology in an interventional study with aged tau transgenic mice. J Neurosci 32(11):3601–3611. doi:10.1523/jneurosci.4922-11.2012
28. Zhang B, Maiti A, Shively S, Lakhani F, McDonald-Jones G, Bruce J, Lee EB, Xie SX, Joyce S, Li C, Toleikis PM, Lee VMY, Trojanowski JQ (2005) Microtubule-binding drugs offset tau sequestration by stabilizing microtubules and reversing fast axonal transport deficits in a tauopathy model. Proc Natl Acad Sci U S A 102(1):227–231
29. Drechsel DN, Hyman AA, Cobb MH, Kirschner MW (1992) Modulation of the dynamic instability of tubulin assembly by the microtubule-associated protein tau. Mol Biol Cell 3(10):1141–1154
30. Wagner U, Utton M, Gallo JM, Miller CC (1996) Cellular phosphorylation of tau by GSK-3 beta influences tau binding to microtubules and microtubule organisation. J Cell Sci 109(6):1537–1543
31. Ksiezak-Reding H, Liu W-K, Yen S-H (1992) Phosphate analysis and dephosphorylation of modified tau associated with paired helical filaments. Brain Res 597(2):209–219. doi:10.1016/0006-8993(92)91476-U

32. Congdon EE, Kim S, Bonchak J, Songrug T, Matzavinos A, Kuret J (2008) Nucleation-dependent tau filament formation: the importance of dimerization and an estimation of elementary rate constants. J Biol Chem 283(20):13806–13816
33. Patrick GN, Zukerberg L, Nikolic M, de la Monte S, Dikkes P, Tsai LH (1999) Conversion of p35 to p25 deregulates Cdk5 activity and promotes neurodegeneration. Nature 402(6762): 615–622. doi:10.1038/45159
34. Dixit R, Ross JL, Goldman YE, Holzbaur EL (2008) Differential regulation of dynein and kinesin motor proteins by tau. Science 319(5866):1086–1089. doi:10.1126/science.1152993
35. Lee VMY, Daughenbaugh R, Trojanowski JQ (1994) Microtubule stabilizing drugs for the treatment of Alzheimer's disease. Neurobiol Aging 15(Suppl 2):S87–S89
36. Brunden KR, Zhang B, Carroll J, Yao Y, Potuzak JS, Hogan AM, Iba M, James MJ, Xie SX, Ballatore C, Smith AB III, Lee VM, Trojanowski JQ (2010) Epothilone D improves microtubule density, axonal integrity, and cognition in a transgenic mouse model of tauopathy. J Neurosci 30(41):13861–13866. doi:10.1523/JNEUROSCI.3059-10.2010, 30/41/13861 [pii]
37. Bollag DM, McQueney PA, Zhu J, Hensens O, Koupal L, Liesch J, Goetz M, Lazarides E, Woods CM (1995) Epothilones, a new class of microtubule-stabilizing agents with a taxol-like mechanism of action. Cancer Res 55(11):2325–2333
38. Prota AE, Bargsten K, Zurwerra D, Field JJ, Díaz JF, Altmann KH, Steinmetz MO (2013) Molecular mechanism of action of microtubule-stabilizing anticancer agents. Science 339(6119):587–590. doi:10.1126/science.1230582
39. Canales A, Nieto L, Rodriguez-Salarichs J, Sanchez-Murcia PA, Coderch C, Cortes-Cabrera-A, Paterson I, Carlomagno T, Gago F, Andreu JM, Altmann KH, Jimenez-Barbero J, Diaz JF (2014) Molecular recognition of epothilones by microtubules and tubulin dimers revealed by biochemical and NMR approaches. ACS Chem Biol 9(4):1033–1043. doi:10.1021/cb400673h
40. Ranade AR, Higgins L, Markowski TW, Glaser N, Kashin D, Bai R, Hong KH, Hamel E, Höfle G, Georg GI (2016) Characterizing the epothilone binding site on β-tubulin by photoaffinity labeling: identification of β-tubulin peptides TARGSQQY and TSRGSQQY as targets of an epothilone photoprobe for polymerized tubulin. J Med Chem 59 (7):3499–3514. doi:10.1021/acs.jmedchem.6b00188
41. Andreu JM, Diaz JF, Gil R, de Pereda JM, Garcia de Lacoba M, Peyrot V, Briand C, Towns-Andrews E, Bordas J (1994) Solution structure of Taxotere-induced microtubules to 3-nm resolution. The change in protofilament number is linked to the binding of the taxol side chain. J Biol Chem 269(50):31785–31792
42. Fellner S, Bauer B, Miller DS, Schaffrik M, Fankhanel M, Spruss T, Bernhardt G, Graeff C, Farber L, Gschaidmeier H, Buschauer A, Fricker G (2002) Transport of paclitaxel (Taxol) across the blood-brain barrier in vitro and in vivo. J Clin Invest 110(9):1309–1318
43. Brunden KR, Yao Y, Potuzak JS, Ferrer NI, Ballatore C, James MJ, Hogan AM, Trojanowski JQ, Smith AB III, Lee VM (2011) The characterization of microtubule-stabilizing drugs as possible therapeutic agents for Alzheimer's disease and related tauopathies. Pharmacol Res 63(4):341–351. doi:10.1016/j.phrs.2010.12.002, S1043-6618(10)00227-6 [pii]
44. Lee JJ, Swain SM (2006) Peripheral neuropathy induced by microtubule-stabilizing agents. J Clin Oncol 24(10):1633–1642. doi:10.1200/jco.2005.04.0543
45. Glaeser H (2011) Importance of P-glycoprotein for drug–drug interactions. In: Fromm FM, Kim BR (eds) Drug transporters. Springer, Berlin, Heidelberg, pp 285–297. doi:10.1007/978-3-642-14541-4_7
46. Cirrito JR, Deane R, Fagan AM, Spinner ML, Parsadanian M, Finn MB, Jiang H, Prior JL, Sagare A, Bales KR, Paul SM, Zlokovic BV, Piwnica-Worms D, Holtzman DM (2005) P-glycoprotein deficiency at the blood-brain barrier increases amyloid-{beta} deposition in an Alzheimer disease mouse model. J Clin Invest 115(11):3285–3290. doi:10.1172/jci25247
47. Ballatore C, Brunden KR, Huryn DM, Trojanowski JQ, Lee VMY, Smith AB III (2012) Microtubule stabilizing agents as potential treatment for Alzheimer's disease and related neurodegenerative tauopathies. J Med Chem 55(21):8979–8996. doi:10.1021/jm301079z

48. Rohena CC, Mooberry SL (2014) Recent progress with microtubule stabilizers: new compounds, binding modes and cellular activities. Nat Prod Rep 31(3):335–355. doi:10.1039/C3NP70092E
49. Conlin A, Fornier M, Hudis C, Kar S, Kirkpatrick P (2007) Ixabepilone. Nat Rev Drug Discov 6(12):953–954
50. Hoffmann J, Fichtner I, Lemm M, Lienau P, Hess-Stumpp H, Rotgeri A, Hofmann B, Klar U (2009) Sagopilone crosses the blood-brain barrier in vivo to inhibit brain tumor growth and metastases. Neuro Oncol 11(2):158–166. doi:10.1215/15228517-2008-072, 15228517-2008-072 [pii]
51. O'Reilly T, Wartmann M, Brueggen J, Allegrini PR, Floersheimer A, Maira M, McSheehy PM (2008) Pharmacokinetic profile of the microtubule stabilizer patupilone in tumor-bearing rodents and comparison of anti-cancer activity with other MTS in vitro and in vivo. Cancer Chemother Pharmacol 62(6):1045–1054. doi:10.1007/s00280-008-0695-9
52. Cartelli D, Casagrande F, Busceti CL, Bucci D, Molinaro G, Traficante A, Passarella D, Giavini E, Pezzoli G, Battaglia G, Cappelletti G (2013) Microtubule alterations occur early in experimental parkinsonism and the microtubule stabilizer epothilone D is neuroprotective. Sci Rep 3:1837. doi:10.1038/srep01837
53. Ruschel J, Hellal F, Flynn KC, Dupraz S, Elliott DA, Tedeschi A, Bates M, Sliwinski C, Brook G, Dobrindt K, Peitz M, Brüstle O, Norenberg MD, Blesch A, Weidner N, Bunge MB, Bixby JL, Bradke F (2015) Systemic administration of epothilone B promotes axon regeneration after spinal cord injury. Science 348(6232):347–352. doi:10.1126/science.aaa2958
54. Distefano M, Scambia G, Ferlini C, Gaggini C, De Vincenzo R, Riva A, Bombardelli E, Ojima I, Fattorossi A, Panici PB, Mancuso S (1997) Anti-proliferative activity of a new class of taxanes (14beta-hydroxy-10-deacetylbaccatin III derivatives) on multidrug-resistance-positive human cancer cells. Int J Cancer 72(5):844–850
55. Metzger-Filho O, Moulin C, de Azambuja E, Ahmad A (2009) Larotaxel: broadening the road with new taxanes. Expert Opin Investig Drugs 18(8):1183–1189
56. Ojima I, Chen J, Sun L, Borella CP, Wang T, Miller ML, Lin S, Geng X, Kuznetsova L, Qu C, Gallager D, Zhao X, Zanardi I, Xia S, Horwitz SB, Mallen-St Clair J, Guerriero JL, Bar-Sagi D, Veith JM, Pera P, Bernacki RJ (2008) Design, synthesis, and biological evaluation of new-generation taxoids. J Med Chem 51(11):3203–3221
57. Mastalerz H, Cook D, Fairchild CR, Hansel S, Johnson W, Kadow JF, Long BH, Rose WC, Tarrant J, Wu MJ, Xue MQ, Zhang G, Zoeckler M, Vyas DM (2003) The discovery of BMS-275183: an orally efficacious novel taxane. Bioorg Med Chem 11(20):4315–4323, S0968089603004954 [pii]
58. Rose WC, Long BH, Fairchild CR, Lee FY, Kadow JF (2001) Preclinical pharmacology of BMS-275183, an orally active taxane. Clin Cancer Res 7(7):2016–2021
59. Sampath D, Discafani CM, Loganzo F, Beyer C, Liu H, Tan X, Musto S, Annable T, Gallagher P, Rios C, Greenberger LM (2003) MAC-321, a novel taxane with greater efficacy than paclitaxel and docetaxel in vitro and in vivo. Mol Cancer Ther 2(9):873–884
60. Bouchet BP, Galmarini CM (2010) Cabazitaxel, a new taxane with favorable properties. Drugs Today (Barc) 46(10):735–742. doi:10.1358/dot.2010.46.10.1519019, 1519019 [pii]
61. Galsky MD, Dritselis A, Kirkpatrick P, Oh WK (2010) Cabazitaxel. Nat Rev Drug Discov 9(9):677–678. doi:10.1038/nrd3254, nrd3254 [pii]
62. Cisternino S, Bourasset F, Archimbaud Y, Semiond D, Sanderink G, Scherrmann JM (2003) Nonlinear accumulation in the brain of the new taxoid TXD258 following saturation of P-glycoprotein at the blood-brain barrier in mice and rats. Br J Pharmacol 138(7):1367–1375. doi:10.1038/sj.bjp.0705150
63. Tang SC, Kort A, Cheung KL, Rosing H, Fukami T, Durmus S, Wagenaar E, Hendrikx JJMA, Nakajima M, van Vlijmen BJM, Beijnen JH, Schinkel AH (2015) P-glycoprotein, CYP3A, and plasma carboxylesterase determine brain disposition and oral availability of the novel taxane cabazitaxel (jevtana) in mice. Mol Pharm 12(10):3714–3723. doi:10.1021/acs.molpharmaceut.5b00470

64. McChesney JD (2012) Design and development of a next generation taxane. Planta Med 78(11):CL18. doi:10.1055/s-0032-1320253
65. Baloglu E, Kingston DG (1999) The taxane diterpenoids. J Nat Prod 62(10):1448–1472
66. Shigemori H, Kobayashi J (2004) Biological activity and chemistry of taxoids from the Japanese yew, Taxus cuspidata. J Nat Prod 67(2):245–256. doi:10.1021/np030346y
67. Fitzgerald DP, Emerson DL, Qian YZ, Anwar T, Liewehr DJ, Steinberg SM, Silberman S, Palmieri D, Steeg PS (2012) TPI-287, a new taxane family member, reduces the brain metastatic colonization of breast cancer cells. Mol Cancer Ther 11(9):1959–1967. doi:10.1158/1535-7163.mct-12-0061
68. Jones ME, Barrett BS, Bell C, Brown E, Feng L, Emerson DL (2007) TPI 287, a third-generation taxane derivative, functionally modulates the MDR1 P-glycoprotein drug transport pump and is active in resistant tumor cells. Mol Cancer Ther 6(12):3399S
69. Jones ME, Bell CB, Schiemann BJ, McChesney JD, Emerson DL (2005) Biological characterization of TPI 287 – a novel third generation taxane analog. Clin Cancer Res 11(24):9089S
70. Defensor EB, Farmer G, Gan L, Boxer A, Shamloo M (2014) (792.14/G4 2014) Effects of TPI 287, a novel taxoid, on a transgenic mouse model of Alzheimer's disease. In: Paper presented at the Neuroscience Meeting Planner, Washington, DC: Society for Neuroscience, Online
71. Pettit GR, Cichacz ZA, Gao F, Boyd MR, Schmidt JM (1994) Isolation and structure of the cancer cell growth inhibitor dictyostatin 1. J Chem Soc, Chem Commun 1111–1112
72. Madiraju C, Edler MC, Hamel E, Raccor BS, Balachandran R, Zhu G, Giuliano KA, Vogt A, Shin Y, Fournier JH, Fukui Y, Bruckner AM, Curran DP, Day BW (2005) Tubulin assembly, taxoid site binding, and cellular effects of the microtubule-stabilizing agent dictyostatin. Biochemistry 44(45):15053–15063
73. Brunden KR, Gardner NM, James MJ, Yao Y, Trojanowski JQ, Lee VMY, Paterson I, Ballatore C, Smith AB III (2013) MT-stabilizer, dictyostatin, exhibits prolonged brain retention and activity: potential therapeutic implications. ACS Med Chem Lett 4(9): 886–889. doi:10.1021/ml400233e
74. Makani V, Zhang B, Han H, Yao Y, Lassalas P, Lou K, Paterson I, Lee VM-Y, Trojanowski JQ, Ballatore C, Smith AB III, Brunden KR (2016) Evaluation of the brain-penetrant microtubule-stabilizing agent, dictyostatin, in the PS19 tau transgenic mouse model of tauopathy. Acta Neuropathol Commun 4:106
75. West LM, Northcote PT, Battershill CN (2000) Peloruside A: a potent cytotoxic macrolide isolated from the New Zealand marine sponge Mycale sp. J Org Chem 65(2):445–449
76. Das V, Miller JH (2012) Microtubule stabilization by peloruside A and paclitaxel rescues degenerating neurons from okadaic acid-induced tau phosphorylation. Eur J Neurosci 35(11): 1705–1717. doi:10.1111/j.1460-9568.2012.08084.x
77. Gaitanos TN, Buey RM, Diaz JF, Northcote PT, Teesdale-Spittle P, Andreu JM, Miller JH (2004) Peloruside A does not bind to the taxoid site on beta-tubulin and retains its activity in multidrug-resistant cell lines. Cancer Res 64(15):5063–5067
78. Huzil JT, Chik JK, Slysz GW, Freedman H, Tuszynski J, Taylor RE, Sackett DL, Schriemer DC (2008) A unique mode of microtubule stabilization induced by Peloruside A. J Mol Biol 378(5):1016–1030
79. Haggarty SJ, Mayer TU, Miyamoto DT, Fathi R, King RW, Mitchison TJ, Schreiber SL (2000) Dissecting cellular processes using small molecules: identification of colchicine-like, taxol-like and other small molecules that perturb mitosis. Chem Biol 7(4):275–286
80. Ayral-Kaloustian S, Zhang N, Beyer C (2009) Cevipabulin (TTI-237): preclinical and clinical results for a novel antimicrotubule agent. Methods Find Exp Clin Pharmacol 31(7): 443–447. doi:10.1358/mf.2009.31.7.1410793, 1410793 [pii]
81. Beyer CF, Zhang N, Hernandez R, Vitale D, Lucas J, Nguyen T, Discafani C, Ayral-Kaloustian S, Gibbons JJ (2008) TTI-237: a novel microtubule-active compound with in vivo antitumor activity. Cancer Res 68(7):2292–2300. doi:10.1158/0008-5472.CAN-07-1420, 68/7/2292 [pii]

82. Beyer CF, Zhang N, Hernandez R, Vitale D, Nguyen T, Ayral-Kaloustian S, Gibbons JJ (2009) The microtubule-active antitumor compound TTI-237 has both paclitaxel-like and vincristine-like properties. Cancer Chemother Pharmacol 64(4):681–689. doi:10.1007/s00280-008-0916-2
83. Zhang N, Ayral-Kaloustian S, Nguyen T, Afragola J, Hernandez R, Lucas J, Gibbons J, Beyer C (2007) Synthesis and SAR of [1,2,4]triazolo[1,5-a]pyrimidines, a class of anticancer agents with a unique mechanism of tubulin inhibition. J Med Chem 50(2):319–327. doi:10.1021/jm060717i
84. Kovalevich J, Cornec AS, Yao Y, James M, Crowe A, Lee VM, Trojanowski JQ, Smith AB III, Ballatore C, Brunden KR (2016) Characterization of brain-penetrant pyrimidine-containing molecules with differential microtubule-stabilizing activities developed as potential therapeutic agents for Alzheimer's disease and related tauopathies. J Pharmacol Exp Ther 357(2): 432–450. doi:10.1124/jpet.115.231175
85. Lou K, Yao Y, Hoye AT, James MJ, Cornec AS, Hyde E, Gay B, Lee VM, Trojanowski JQ, Smith AB III, Brunden KR, Ballatore C (2014) Brain-penetrant, orally bioavailable microtubule-stabilizing small molecules are potential candidate therapeutics for Alzheimer's disease and related tauopathies. J Med Chem 57(14):6116–6127. doi:10.1021/jm5005623
86. Cornec AS, James MJ, Kovalevich J, Trojanowski JQ, Lee VM, Smith AB III, Ballatore C, Brunden KR (2015) Pharmacokinetic, pharmacodynamic and metabolic characterization of a brain retentive microtubule (MT)-stabilizing triazolopyrimidine. Bioorg Med Chem Lett 25(21):4980–4982. doi:10.1016/j.bmcl.2015.03.002
87. Zhang N, Ayral-Kaloustian S, Nguyen T, Hernandez R, Beyer C (2007) 2-cyanoaminopyrimidines as a class of antitumor agents that promote tubulin polymerization. Bioorg Med Chem Lett 17(11):3003–3005. doi:10.1016/j.bmcl.2007.03.070, S0960-894X(07)00373-3 [pii]
88. Zhang N, Ayral-Kaloustian S, Nguyen T, Hernandez R, Lucas J, Discafani C, Beyer C (2009) Synthesis and SAR of 6-chloro-4-fluoroalkylamino-2-heteroaryl-5-(substituted)phenylpyrimidines as anti-cancer agents. Bioorg Med Chem 17(1):111–118. doi:10.1016/j.bmc.2008.11.016, S0968-0896(08)01081-X [pii]
89. Lamberth C, Trah S, Wendeborn S, Dumeunier R, Courbot M, Godwin J, Schneiter P (2012) Synthesis and fungicidal activity of tubulin polymerisation promoters. Part 2: Pyridazines. Bioorg Med Chem 20(9):2803–2810. doi:10.1016/j.bmc.2012.03.035, S0968-0896(12)00225-8 [pii]
90. Crowley PJ, Lamberth C, Müller U, Wendeborn S, Nebel K, Williams J, Sageot O-A, Carter N, Mathie T, Kempf H-J, Godwin J, Schneiter P, Dobler MR (2010) Synthesis and fungicidal activity of tubulin polymerisation promoters. Part 1: Pyrido[2,3-b]pyrazines. Pest Manag Sci 66:178–185
91. Lamberth C, Dumeunier R, Trah S, Wendeborn S, Godwin J, Schneiter P, Corran A (2013) Synthesis and fungicidal activity of tubulin polymerisation promoters. Part 3: Imidazoles. Bioorg Med Chem 21(1):127–134. doi:10.1016/j.bmc.2012.10.052
92. Dumeunier R, Lamberth C, Trah S (2013) Synthesis of tetrasubstituted pyrazoles through different cyclization strategies; isosteres of imidazole fungicides. Synlett 24(9):1150–1154. doi:10.1055/s-0033-1338433
93. Taggi AE, Stevenson TM, Bereznak JF, Sharpe PL, Gutteridge S, Forman R, Bisaha JJ, Cordova D, Crompton M, Geist L, Kovacs P, Marshall E, Sheth R, Stavis C, Tseng C-P (2016) Tubulin modulating antifungal and antiproliferative pyrazinone derivatives. Bioorg Med Chem 24(3):435–443. doi:10.1016/j.bmc.2015.08.038
94. Klar U, Graf H, Schenk O, Röhr B, Schulz H (1998) New synthetic inhibitors of microtubule depolymerization. Bioorg Med Chem Lett 8(11):1397–1402. doi:10.1016/S0960-894X(98)00226-1
95. Shintani Y, Tanaka T, Nozaki Y (1997) GS-164, a small synthetic compound, stimulates tubulin polymerization by a similar mechanism to that of Taxol. Cancer Chemother Pharmacol 40(6):513–520

96. Pandit B, Hu Z, Chettiar SN, Zink J, Xiao Z, Etter JP, Bhasin D, Li P-K (2013) Structure–activity relationship studies of thalidomide analogs with a taxol-like mode of action. Bioorg Med Chem Lett 23(24):6902–6904. doi:10.1016/j.bmcl.2013.09.084
97. Li PK, Pandit B, Sackett DL, Hu Z, Zink J, Zhi J, Freeman D, Robey RW, Werbovetz K, Lewis A, Li C (2006) A thalidomide analogue with in vitro antiproliferative, antimitotic, and microtubule-stabilizing activities. Mol Cancer Ther 5(2):450–456. doi:10.1158/1535-7163. MCT-05-0254, 5/2/450 [pii]
98. Marks MG, Shi J, Fry MO, Xiao Z, Trzyna M, Pokala V, Ihnat MA, Li P-K (2002) Effects of putative hydroxylated thalidomide metabolites on blood vessel density in the chorioallantoic membrane (CAM) assay and on tumor and endothelial cell proliferation. Biol Pharm Bull 25(5):597–604. doi:10.1248/bpb.25.597
99. Inatsuki S, Noguchi T, Miyachi H, Oda S, Iguchi T, Kizaki M, Hashimoto Y, Kobayashi H (2005) Tubulin-polymerization inhibitors derived from thalidomide. Bioorg Med Chem Lett 15(2):321–325. doi:10.1016/j.bmcl.2004.10.072
100. Rashid A, Kuppa A, Kunwar A, Panda D (2015) Thalidomide (5HPP-33) suppresses microtubule dynamics and depolymerizes the microtubule network by binding at the vinblastine binding site on tubulin. Biochemistry 54(12):2149–2159. doi:10.1021/bi501429j
101. Wang Z, Yang D, Mohanakrishnan AK, Fanwick PE, Nampoothiri P, Hamel E, Cushman M (2000) Synthesis of B-ring homologated estradiol analogues that modulate tubulin polymerization and microtubule stability. J Med Chem 43(12):2419–2429. doi:10.1021/jm0001119
102. D'Amato RJ, Lin CM, Flynn E, Folkman J, Hamel E (1994) 2-Methoxyestradiol, an endogenous mammalian metabolite, inhibits tubulin polymerization by interacting at the colchicine site. Proc Natl Acad Sci U S A 91(9):3964–3968
103. Verdier-Pinard P, Wang Z, Mohanakrishnan AK, Cushman M, Hamel E (2000) A steroid derivative with paclitaxel-like effects on tubulin polymerization. Mol Pharmacol 57(3):568–575. doi:10.1124/mol.57.3.568
104. Reddy MV, Akula B, Cosenza SC, Lee CM, Mallireddigari MR, Pallela VR, Subbaiah DR, Udofa A, Reddy EP (2012) (Z)-1-aryl-3-arylamino-2-propen-1-ones, highly active stimulators of tubulin polymerization: synthesis, structure-activity relationship (SAR), tubulin polymerization, and cell growth inhibition studies. J Med Chem 55(11):5174–5187. doi:10.1021/jm300176j
105. Kamal A, Reddy VS, Shaik AB, Kumar GB, Vishnuvardhan MVPS, Polepalli S, Jain N (2015) Synthesis of (Z)-(arylamino)-pyrazolyl/isoxazolyl-2-propenones as tubulin targeting anticancer agents and apoptotic inducers. Org Biomol Chem 13(11):3416–3431. doi:10.1039/C4OB02449D
106. Yang WS, Shimada K, Delva D, Patel M, Ode E, Skouta R, Stockwell BR (2012) Identification of simple compounds with microtubule-binding activity that inhibit cancer cell growth with high potency. ACS Med Chem Lett 3(1):35–38. doi:10.1021/ml200195s
107. Pees K-J, Albert G (1993) Triazolopyrimidine derivatives with fungicidal activity. European Patent 550113
108. Brunden KR, Trojanowski JQ, Smith AB III, Lee VM, Ballatore C (2013) Microtubule-stabilizing agents as potential therapeutics for neurodegenerative disease. Bioorg Med Chem 22(18):5040–5049. doi:10.1016/j.bmc.2013.12.046

Top Med Chem (2017) 24: 181–198
DOI: 10.1007/7355_2016_18

Published online: 4 March 2017

PET Imaging Agents for Alzheimer's Disease

Seok Rye Choi, Karl Ploessl, Lin Zhu, and Hank F. Kung

Abstract In the past four decades, PET (positron emission tomography) imaging has been developed into a common diagnostic tool. After an i.v. injection of a small amount of radioactive probe, PET images are taken based on the decay of short-lived positron-emitting isotopes. The emitted positron collides with a neighboring electron to produce two 511 KeV gamma rays roughly 180 apart. It is based on this physical principal and the positron coincident signals that the point source of the gamma rays can be deduced by using a coincident circuit for detectors 180 apart. Currently, there are three FDA-approved imaging agents (Amyvid, NeuraCeq, and Vizamyl) available for mapping Aβ in the brain of patients suspected of having Alzheimer's disease. There is a strong interest in measuring tau deposition by PET imaging in a quantitative manner. Several useful Tau imaging agents have been tested in humans; however, they have not been fully validated as those of Aβ imaging agents.

S.R. Choi and K. Ploessl
Department of Radiology, University of Pennsylvania, Philadelphia, PA 19104, USA

Five Eleven Pharma Inc., Philadelphia, PA 19104, USA

L. Zhu
Department of Radiology, University of Pennsylvania, Philadelphia, PA 19104, USA

Institute for Brain Disorders, Capital Medical University, Beijing, China

Key Laboratory of Radiopharmaceuticals, Beijing Normal University, Ministry of Education, Beijing 100875, P.R. China

H.F. Kung (✉)
Department of Radiology, University of Pennsylvania, Philadelphia, PA 19104, USA

Five Eleven Pharma Inc., Philadelphia, PA 19104, USA

Institute for Brain Disorders, Capital Medical University, Beijing, China
e-mail: kunghf@sunmac.spect.upenn.edu

Keywords Alzheimer's disease, Amyloid plaques, Imaging agent, Positron emission tomography, Tau aggregates

Contents

1 Introduction

1.1 In Vivo Positron Emission Tomography Imaging

In the past 4 decades, PET (positron emission tomography) imaging has been developed into a common diagnostic tool. After an i.v. injection of a small amount of radioactive probe, PET images are taken based on the decay of short-lived positron-emitting isotopes. The emitted positron collides with a neighboring electron to produce two 511 KeV gamma rays roughly 180° apart. It is based on this physical principal and the positron coincident signals that the point source of the gamma rays can be deduced, by using a coincident circuit for detectors 180° apart [1].

Two most commonly used positron-emitting isotopes are ^{11}C ($T_{1/2} = 20$ min) and ^{18}F ($T_{1/2} = 110$ min) (Table 1) [1]. A metabolic tracer for brain imaging, ^{11}C glucose, has been reported for studying glucose utilization in the brain

Table 1 Radionuclides for PET imaging agents

Radionuclides	$T_{1/2}$	Production methods[a]
^{11}C	20 min	N-14 (p, α) C-11, cyclotron
^{18}F	110 min	O-18 (p, n) F-18, cyclotron
^{13}N	10 min	O-16 (p, α) N-13, cyclotron
^{68}Ga	68 min	^{68}Ge/^{68}Ga Generator

[a]In producing isotopes in this list, a proton, p, is accelerated in a cyclotron and when the high energy beam hits the target containing the parent element it knocks off either an alpha particle, α, or a neutron, n, from the parent isotope to produce the daughter isotope listed above

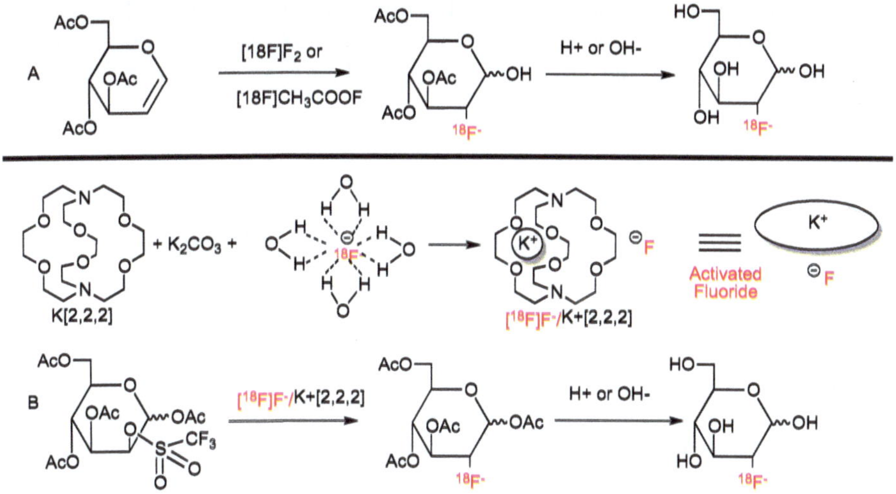

Fig. 1 The schemes show the production of [^{18}F]FDG (2-fluoro-2-deoxy-D-glucose). (**a**) The original method for preparation of [^{18}F]FDG employed fluorine gas ([^{18}F]F_2) [4, 5]. (**b**) The improved synthesis of [^{18}F]FDG uses [^{18}F]fluoride ion, potassium carbonate, and K[2.2.2], a cryptand, to activate the fluoride ion [6, 7]. This nucleophilic reaction improved production of [^{18}F]FDG

[2, 3]. However, due to its limited physical half-life, it is challenging to utilize ^{11}C radiopharmaceuticals. Only a few major medical centers and clinics with technical skills can make ^{11}C radiolabeled probes for patient studies. The half-life of ^{18}F (110 min) is 5.5 times longer than that of ^{11}C ($T_{1/2} = 20$ min). Thus, it is generally accepted that ^{18}F tracers are more practical. An off-site cyclotron can manufacture them commercially. The distribution can be centralized by a radiopharmacy. Originally, [^{18}F]FDG (2-fluoro-2-deoxyglucose) was prepared with fluorine gas ([^{18}F]F_2; Fig. 1a). The process is rather complicated, and the labeling yield was less than 30%. The reaction has been improved significantly by a direct nucleophilic reaction using a fluoride ion, potassium carbonate, and K[2.2.2] Kryptofix®, a cryptand, to activate the fluoride ion (see Fig. 1a, b) [4, 6]. This procedure can produce >370 GBq (>10 Ci) of [^{18}F]FDG per batch (sufficient for hundreds of patient doses).

Currently, [^{18}F]FDG, a PET imaging agent with widespread clinical application, is transported across cell membranes via a glucose transporter and then phosphorylated by hexokinase in the cytosol. The process is closely linked with cellular metabolism, including glucose usage. FDG as an imaging agent is commonly employed for studying glucose metabolism in the brain as well as in detecting tumors, which show a higher rate of metabolism [8]. FDG-PET provides a diagram of cancer cell activity within the body. PET-CT (computer tomography) also offers a fusion image combining regional glucose metabolism and anatomical structure in one imaging session. This combined imaging method is now the standard of care in oncology for diagnosis, staging, and monitoring therapy [9].

PET imaging in conjunction with FDG has been approved by the FDA for diagnosis of Alzheimer's disease [10, 11], but the FDG regional distribution in the brain is a reflection of glucose metabolism, which may not be directly associated with AD symptoms or disease progression.

2 Alzheimer's Disease (AD): Pathogenesis and Amyloid Imaging

Alzheimer's disease (AD) is a neurodegenerative disorder of the brain, characterized by a loss of mental function and progressive dementia. This disease is important because it affects millions of older patients. Clinical symptoms of AD include cognitive decline, irreversible memory loss, disorientation, and language impairment. Major neuropathological observations of postmortem AD brain include the presence of senile plaques containing β-amyloid (Aβ) aggregates and neurofibrillary tangles (NFTs) containing highly phosphorylated tau proteins (Fig. 2a). Significant evidence suggests that accumulation and aggregation of Aβ peptides may play a major role in AD pathogenesis [13, 16, 17]. The aggregated Aβ, produced by proteolytic processing of the β-amyloid precursor protein (Fig. 2b), may start a series of events (Fig. 2a).

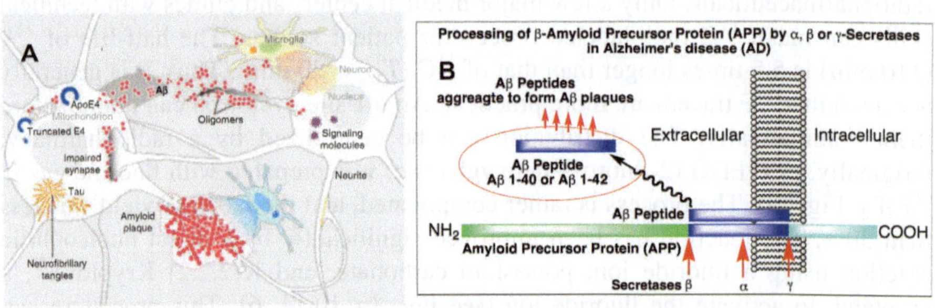

Fig. 2 (**a**) Processes implicated in AD pathogenesis. Aβ peptides produced by neurons aggregate into a variety of assemblies (e.g., oligomers, plaques), some of which may impair synapses and neuronal dendrites. Build-up of pathogenic Aβ aggregates could result from increased production or from deficient clearance mechanisms. This multifactorial scenario leads to progressive disintegration of neural circuits, isolation and loss of neurons, network failure, and neurological decline (reproduced with permission from Science, TBD) [12]. (**b**) A simplified model of the excessive Aβ production leading to Alzheimer's disease (AD). Normally, the β-amyloid precursor protein (APP) is metabolized by at least three proteases (α-, β-, and γ-secretases). Aβ peptides are produced by β-secretase, forming the N-terminus, and by γ-secretase, forming the C-terminus of Aβ. The fibrillar aggregates of amyloid peptides, Aβ40 and Aβ42, are the major constituent of senile plaques in AD patients. The Aβ aggregates are believed to be responsible for initiating a cascade of events leading to neurotoxicity and AD [13–15]

2.1 *Amyloid (Aβ) Plaque Imaging Agents*

Development of PET imaging agents targeting Aβ aggregates in the living human brain has been a research topic for many studies and reviews [18–22]. An essential prerequisite for a successful imaging agent for Aβ plaques in the brain is the ability to rapidly penetrate the intact blood–brain barrier. A small and moderately lipophilic PET agent exhibiting an excellent initial brain uptake is an imperative requirement. Not only should the agent have a high initial brain uptake, but also have fast washout kinetics from the normal brain. The specific to nonspecific binding of the PET signal will be critical for quantitative determination of Aβ plaques in the brain [23–26]. These are highly desirable properties for a useful brain-imaging agent targeting Aβ plaques. The development of diagnostic imaging agents targeting Aβ aggregates may lead to improved selection and monitoring of patients undergoing drug treatment designed to reverse the build-up of Aβ plaque in the brain [16].

Thousands of compounds have been prepared and tested for imaging Aβ aggregates. However, the clinically validated and FDA-approved Aβ plaque-targeting PET imaging agents can generally be classified into two main groups: stilbenes and benzothiazoles (Fig. 3a, b, respectively). Chemical structures of benzothiazoles (PIB analogs [19]) and stilbenes (SB-13 analogs [21]) are similar; both are highly conjugated aromatic and planar molecules – an important attribute for insertion between the β-sheets of Aβ aggregates. Both types of molecules contain a common electron donating group (*N*-methylamine) at the end of the benzene ring [27], and they compete for similar binding sites on the Aβ aggregates [19–22]. It is important to note that these PET imaging agents for Aβ aggregates not only have fulfilled the

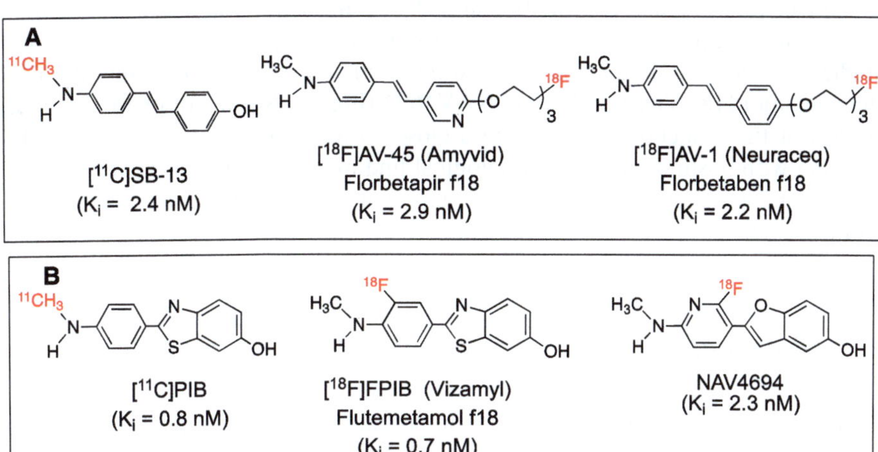

Fig. 3 Chemical structures of ^{11}C and ^{18}F labeled amyloid imaging agents. All of these agents have been successfully tested in humans. The FDA approved the commercial use of florbetapir f18 (Amyvid™) in 2012, flutemetamol f18 (Vizamyl™) in 2013, and florbetaben f18 (NeuraCeq™) in 2014; the FDA approval of NAV4694 is still pending

basic requirements as listed above, they also have been extensively tested in living AD human brain. Postmortem studies have confirmed that the extent of Aβ aggregate accumulation in the brain (measured by pathological examination – plaque counting) correlated with the in vivo PET images. This is a very high criterion. Because PET imaging prior to death and the subsequent postmortem AD human brain studies involving hundreds of patients is very expensive, such studies could only be carried out by commercial companies with the intention to submit a new drug application (NDA).

In 2001, early reports on radiolabeled stilbenes (Fig. 3a) showed high binding affinity and selectivity towards Aβ aggregates, and the autoradiography of postmortem human AD brain sections confirmed the binding [27]. From these stilbenes, [^{11}C]SB-13 was selected and successfully tested in normal controls and AD patients [28]. Two successive F-18 labeled agents, NeuraCeq™ (florbetaben f18, AV-1, developed by Schering/Piramal Health) [29] and Amyvid™ (florbetapir f18, AV-45, developed by Avid/Eli Lilly) [30], were approved by the FDA for commercial use for detecting amyloid accumulation in the brain. The structure–activity relationship for this series of agents has been reviewed [20, 21]. Among the benzothiazole-based agents (Fig. 3b), [^{11}C]PIB is now the agent of choice for C-11 PET imaging, and a wealth of information on its use in AD patients has been reported [31–33]. The benzothiazoles as PET imaging agents for detecting the presence of amyloid accumulations in the brain have been reviewed [19, 22, 23]. FDA has approved the use of Vizamyl™ (flutemetamol f18, FPIB, developed by GE Health) in detecting Aβ plaques in the brain. Flutemetamol PET imaging of brain beta-amyloid plaque density improves diagnostic accuracy in cognitively impaired patients [34]. Another benzothiazole-related agent, NAV4694 (originally as AZ4694, developed by AstraZeneca/Navidea), was also comprehensively tested in humans [35, 36], and it is likely that NDA submission to the FDA is pending.

In summary, four F-18 PET imaging agents targeting Aβ aggregates have been studied in humans: florbetapir f18 (Amyvid™), florbetaben f18 (NeuraCeq™), flutemetamol f18 (Vizamyl™), and NAV4694 (Fig. 3). Results of phase 3 clinical studies of these agents demonstrated a close correlation of in vivo amyloid PET imaging with postmortem histopathological findings. The FDA has approved three agents for routine clinical uses, and the approval of the fourth agent, NAV4694, is still pending.

2.2 *Potential Applications of Imaging Plaques in the Brain*

Clinically, β-amyloid imaging will be important for the differential diagnosis of dementia. A negative scan will indicate that dementia is not related to AD. A positive scan will imply a higher risk for AD. It is also important to detect preclinical AD and predict the progression from mild cognitive impairment (MCI) to AD, thereby establishing a temporal correlation between the β-amyloid imaging and neurological status. Early monitoring of patients with genetic or family risk factors will also be highly desirable and reports of these type of early monitoring are coming out [37]. One of the significant considerations is for ApoE4-positive subjects. It has been demonstrated that ApoE4 carriers will show early development of positive scans and progression of plaque-build-up. Aβ plaque-positive patients even without memory deficit are likely at risk for AD [38]. PET imaging studies in patients with risk factors suggest that brain amyloid deposition is bad news, though it can take years until cognitive consequences emerge, and the initial cognitive decrements may be subtle.

Many treatments for AD have been aimed at stopping the formation of, or reversing, the Aβ aggregates. The two major approaches of the antiplaque treatments have been the inhibition of β- or γ-secretase (reducing Aβ production) and immunotherapy (removing Aβ from the brain). The past few years have witnessed a series of failures in AD drug development [39–41]. Antiplaque treatments based on the inhibition of γ- and β-secretase (also known as BACE) by small molecules have had difficulty meeting clinical endpoints. However, active efforts in this area of drug development are ongoing [42–46]. Negative clinical results have also been reported for the anti-amyloid antibodies solanezumab (Eli Lilly) and bapineuzumab (Pfizer), two leading immunotherapy drugs designed to remove Aβ in the brain. Additional trials are currently ongoing to further explore the potential benefits. The NIH has recently announced the selection of crenezumab, an antibody by Genentech, as the first therapeutic drug to be evaluated in a preventative clinical trial. Some recently developed amyloid plaque imaging agents, particularly Amyvid™, may be useful to select patients, at the early phase of amyloid plaque development.

Human amyloid imaging will be useful for evaluation of risk factors associated with Alzheimer's disease. Although this imaging procedure on determining the presence or the absence of Aβ plaques (or Tau, see Sect. 3 in this chapter) in the brain may prove to be tremendously valuable once it is incorporated into the routine clinical work-up of dementia patients, Aβ plaque imaging is not without its critics. The debate on the utility of human amyloid imaging may be just a part of the validation process for this new technology. The amyloid-specific imaging will provide a new tool in selecting and monitoring patients enrolled in clinical trials for AD drugs specifically aiming to reduce Aβ-amyloid in the brain. Correlation of plaque load vs PET imaging has been established via phase III clinical studies of three FDA-approved imaging agents, florbetapir f18 (Amyvid™), flutemetamol f18 (Vizamyl™), and florbetaben f18 (NeuraCeq™) [30, 34, 47]. Plaque load was not directly correlated with AD symptom, only a reflection of a risk factor. Using PET Aβ plaque imaging for selection of patients for drug trial evaluation of the effectiveness of anti-Aβ plaque treatment starts with selecting suitable AD patients with Aβ plaques.

3 Tau Imaging Agents

3.1 Tau Aggregates as a Risk Factor for Developing Alzheimer's Disease

The presence of tau aggregates in the brain is a major risk factor for the development of AD [48–50]. Normally, tau proteins serve several essential functions inside neurons, including promoting the assembly of tubulin into microtubules and the stabilization of their structure. Tau is mainly found in the axons of the CNS and is comprised of six isoforms generated by alternative splicing; the longest of them has 441 amino acids. There are three (3R tau) or four (4R tau) microtubule-binding repeats. The hyperphosphorylation of the tau protein at several serine, threonine, and tyrosine sites usually precedes both conformational changes and cleavage of the protein. Tau and mutated tau contain many potential cleavage positions vulnerable to multiple proteases, leading to breakdown products that are believed to be toxic and more susceptible to aggregation. Other biochemical modifications of tau proteins are oxidation, ubiquitination, O-GlcNAc glycosylation, and acetylation, the last of which impairs microtubule binding of tau, promotes aggregation, and potentially interferes with ubiquitination that leads to the degradation of tau. Recent reports suggest that the intracellular transport of the tau protein is consistent with tau protein diffusion along the microtubule lattice [17, 51, 52]. In addition to Alzheimer's disease, the aggregation of tau is also a presumed cause of several diseases known as tauopathies [53–55]. These tauopathies are heterogeneous, with spatiotemporal differences in affected brain areas and also differences in the spliced variants of aggregated tau as "paired helical filaments" (PHFs) [56]. The presence of tau aggregates in the absence of deposits of Aβ peptides ("tau only" brain) is seen in several tauopathies, such as Pick's disease, progressive supranuclear palsy (PSP), and corticobasal degeneration. Frontotemporal lobar degeneration (FTD) is the most common form of tauopathies.

Various approaches to the development of tau therapeutics have also been reported [57, 58]. However, similar to anti-amyloid treatment, it will be essential to select patients specifically for tau aggregates in the brain. This approach is further complicated by the presence of various forms of modified or mutated tau aggregates, which may contain different binding sites. Heterogeneous binding sites would create a challenging problem for the development of PET imaging agents for tau aggregates.

3.2 Imaging Agents Tested for Tau Aggregates in Alzheimer's Disease

Because the accumulation of tau NFTs occurs in AD, and the deposition of Tau aggregates correlates well with the progression of symptoms (cognitive impairment) in AD patients [49, 50, 59], there is an urgent need for tau-specific probes for detecting and monitoring the progression of AD. There have been many attempts to develop tau-specific imaging agents, mainly for PET imaging.

3.2.1 FDDNP

The first notable agent was [^{18}F]FDDNP (Fig. 4) [60–62]. FDDNP is a PET imaging probe with binding affinity for "amyloid and amyloid-like structures." It was produced for in vivo brain imaging of patients with AD and tauopathies (see discussion below) [60, 63, 64]. It was reported that this agent is useful for binding to both Aβ plaques and tau. The signal-to-noise ratios in humans were reported to be distinctively related to Aβ plaques and tau. The multiple binding sites for this probe make the interpretation of PET imaging highly puzzling – binding to both Aβ plaques and tau in the brain regions may not be easily and simultaneously validated (Fig. 4).

3.2.2 5H-pyrido[4,3-b]indoles (T807/AV1451 and T808) as Tau Imaging Agents

The Siemens molecular imaging group, which has reported a series of tau imaging probes (such as T807 and T808, Fig. 5), appears to have made the most progress in this field [65]. In the past 10 years, multiple patent applications have appeared in the literature. T807 and T808, probes for tau imaging, have been sold to Avid/Lilly, and T807 is now designated as AV1451 [65–69]. The leading candidate for tau imaging is ^{18}F-AV1451/T807 with a core structure of 5H-pyrido[4,3-b]indoles (Fig. 5). In in vitro binding assays, using tau-enriched AD brain sections with autoradiography, the K_d for AV1451/T807 was 14.6 nM and Aβ aggregate binding was >2,000 nM [71]; therefore, the selectivity for tau binding is excellent. Human imaging studies of this leading tau imaging candidate have been reported [66, 68, 71]. ^{18}F-AV1451/T807 PET imaging could be valuable as a biomarker that reflects both the progression of AD tauopathy and the emergence of clinical impairment [68]. Preliminary

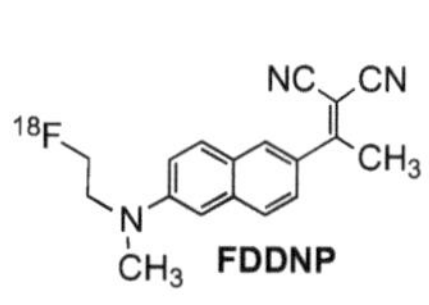

Fig. 4 FDDNP as a tau imaging agent

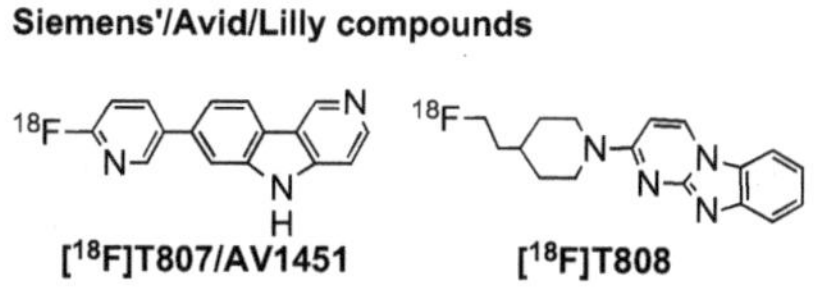

Fig. 5 T807/AV1451 and T808 as tau imaging agents

results of regional and substrate-specific autoradiographic and in vitro binding patterns of ^{18}F-AV1451/T807 in postmortem human brain sections suggested that it might be a useful surrogate marker for detection of tau in living human brain [72].

3.2.3 PBB3 as Tau Imaging Agents

Previously, a series of 2-ethenyl-benzoxazole derivatives (BF compounds; Fig. 6) was reported as Aβ and tau imaging agents. However, the binding selectivity of the series of probes was not optimal, as the affinity for tau fibrils was similar to that of Aβ fibrils [73–75]. Recent publications of using ^{11}C-PBB3 for imaging tau suggest that it may serve as a biomarker for detecting tau in the brain [76, 77]. Despite its brilliant fluorescent signal, the drawbacks of PBB3 are related to its low brain uptake after an i.v. injection. Although new PET image analysis methods have been developed for this probe, the low uptake in the brain makes quantification much harder to accomplish [70, 76, 78].

3.2.4 Quinolines (THK Series) as Tau Imaging Agents

More recently, quinoline-based probes (Fig. 7) for imaging tau pathology have been reported (from the research group of Tohoku University) [82–84]. It was reported that [^{18}F]THK-5105 appeared to show a higher affinity for tau fibrils than for Aβ aggregates (K_d for tau fibrils 1.99 nM vs 30.3 nM for Aβ aggregates). In vitro autoradiography studies of postmortem brain samples suggested different

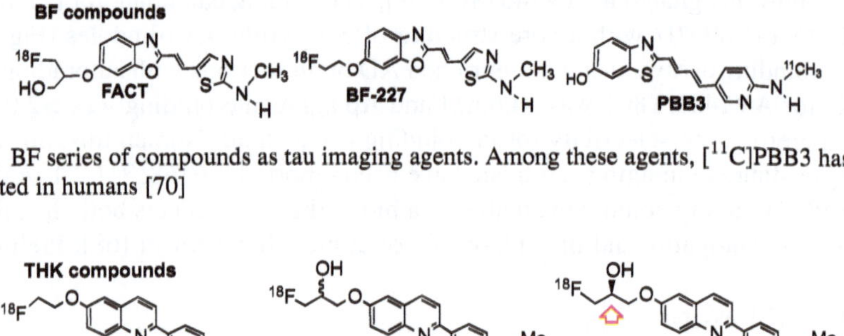

Fig. 6 BF series of compounds as tau imaging agents. Among these agents, [^{11}C]PBB3 has been evaluated in humans [70]

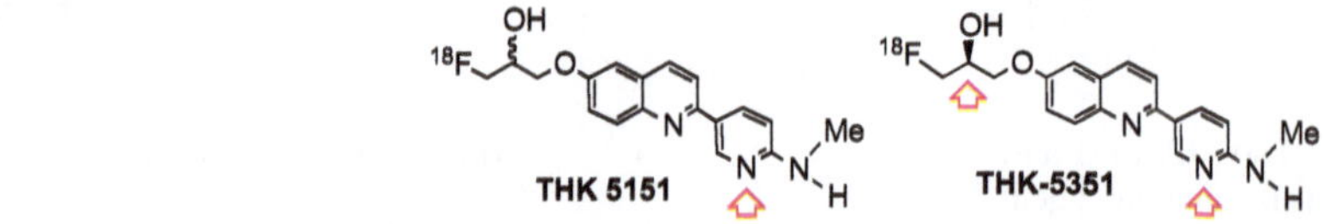

Fig. 7 THK compounds as tau imaging agents. It is important to note that THK5117 (phenyl) and THK5151 (pyridinyl) are two core similar structures for tau binding; their *S*-isomers, THK5317 [79] and THK5351 [80], both displayed high binding affinity and selectivity [81]

distribution patterns between THK-523 and PIB, showing different binding between Aβ and tau. These results suggest that the THK ligands are reasonable candidates for testing in humans [85]. The *S*-isomers of THK-5117, called THK-5317 (Fig. 7), were prepared and displayed better target-to-nontarget ratios [80, 82]. Human PET imaging displayed high selective binding to tau [80]. In a related effort, a stereo-selective quinolone-pyridyl derivative, ^{18}F-THK5351 (the *S*-isomer of THK5151), was prepared and tested in humans [79, 81]. The main difference in uptake and kinetics between THK-5317 and THK-5351 is likely due to the additional nitrogen atom (pyridyl), which makes the agent less lipophilic. Therefore, the in vivo kinetics in the brain were faster and provided a good selective tau signal [81].

3.2.5 Other Tau Imaging Agents

As mentioned above, there are likely multiple mutated variants of tau aggregates and different binding sites. It is likely that many other tau imaging agents targeting different sites could be developed.

It was reported that lansoprazole and astemizole (Fig. 8) have nanomolar affinity for certain forms of tau and selectivity for tau over Aβ plaques [86]. These agents are normally for the treatment of allergies and gastrointestinal disorders, respectively. Recently, ^{11}C and ^{18}F labeled *N*-methyl lansoprazole have been reported as tau imaging agents [87, 88]. In vivo studies suggested a distinctive regional brain distribution different from that of Aβ plaques, but the binding did not overlap with the other tau agents listed above. Additional tau targeting probe, MK-6240, was recently reported from Merck [89], displaying very high affinity ($K_i = 0.36 \pm 0.8$ nM) and selectivity in NFT-rich AD brain homogenates. In vitro autoradiography also showed regional AD brain distribution consistent with known tau (NFT) burden.

A recent publication evaluating binding sites for tau imaging agents [90] suggested that there are two distinctive binding pockets for [^{18}F]THK-523 and [^{18}F]T807 (AV1451) using AD human brain homogenates. It is possible that we are entering a new phase of developing PET imaging agent specifically targeting tau aggregates. Many different binding sites may be available within the β sheet structure of tau aggregates, and the binding of each agent may show different preferences to different pockets.

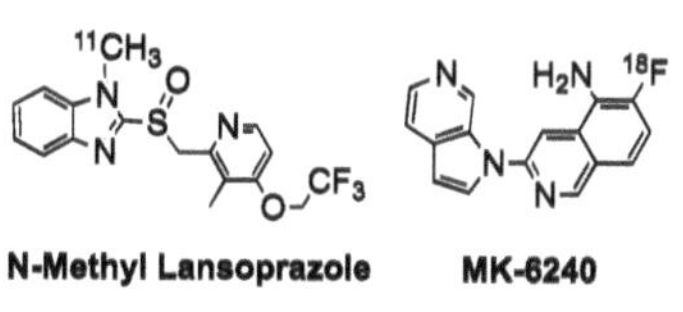

Fig. 8 Other tau imaging agents: *N*-methyl-lansoprazole and MK-6240

3.3 Summary for Tau Imaging Agents

It is generally believed that tau pathology is closely linked to symptoms and progression of AD, and there is a strong interest in measuring tau deposition by PET imaging in a quantitative manner. Several useful Tau imaging agents have been tested in humans; however, they have not been fully validated as those of Aβ imaging agents. Correlation of tau deposition vs. PET images may require additional systematic studies, perhaps by a series of brain autopsies of a full range of AD patients, to fully demonstrate the effectiveness of monitoring tau accumulation in patients by PET imaging.

4 Conclusions and Perspective

Currently, there are three FDA-approved imaging agents (Amyvid, NeuraCeq, and Vizamyl) available for mapping Aβ in the brain of patients suspected of having Alzheimer's disease. Although the Aβ imaging in the brain is available, it has not yet achieved widespread use due to a limited Medicare reimbursement. When successful treatments of Alzheimer's disease are developed, its use may increase in the future. More and more tau targeting agents for PET imaging have been developed and clinical studies have generated exciting data suggesting the importance of tau in the disease process. However, there is still uncertainty with the correlation of PET imaging signal vs the presence of tau (or more precisely which version of mutated tau aggregates). Off-target signals also make the interpretation of the brain images difficult. It is likely that clinical studies correlating the tau signal with postmortem human brain sections will be part of the answers. The build-up of Aβ and tau and their relationship with disease process will likely be elucidated by PET imaging. The information will also play an important role in development of drug treatments.

In addition to the imaging agents based on Aβ and tau, many other PET imaging agents have been proposed for studying the brain function of AD patients, such as agents for neuroinflammation, nicotinic receptors, serotonin receptors, etc. So far none has gone through validation steps which are necessary to correlate the in vivo binding and the level of binding sites in connection with the disease status. Further work will be needed in order to improve tau binding and selectivity at various binding pocket sites.

References

1. Cherry S, Sorenson J, Phelps M (2003) Physics in nuclear medicine, 3rd edn. W.B. Saunders, New York

2. Phelps ME, Mazziotta JC, Schelbert HR (1986) Positron emission tomography and autoradiography: principles and applications for the brain and heart. Raven, New York
3. Sokoloff L (1986) Positron emission tomography and autoradiography: principles and applications for the brain and heart. Raven, New York
4. Fowler JS, MacGregor RR, Wolf AP, Farrell AA, Karlstrom KI, Ruth TJ (1981) A shielded synthesis system for production of 2-deoxy-2-[18F]fluoro-D-glucose. J Nucl Med 22:376–380
5. Ido T, Wan CN, Fowler JS, Wolf AP (1977) Fluorination with F2: a convenient synthesis of 2-deoxy-2-fluoro-D-glucose. J Org Chem 42:2341
6. Hamacher K, Coenen HH, Stocklin G (1986) Efficient stereospecific synthesis of no-carrier-added 2-[18F]-fluoro-2-deoxy-D-glucose using aminopolyether supported nucleophilic substitution. J Nucl Med 27:235–238
7. Alexoff DL, Casati R, Fowler JS, Wolf AP, Shea C, Schlyer DJ et al (1992) Ion chromatographic analysis of high specific activity 18FDG preparations and detection of the chemical impurity 2-deoxy-2-chloro-D-glucose. Int J Rad Appl Instrum A 43:1313–1322
8. Gambhir SS (2008) Molecular imaging of cancer: from molecules to humans. Introduction. J Nucl Med 49(Suppl 2):1S–4S
9. Zhu L, Ploessl K, Kung HF (2013) Expanding the scope of fluorine tags for PET imaging. Science 342:429–430
10. Small GW, Ercoli LM, Silverman DH, Huang SC, Komo S, Bookheimer SY et al (2000) Cerebral metabolic and cognitive decline in persons at genetic risk for Alzheimer's disease. Proc Natl Acad Sci U S A 97:6037–6042
11. Silverman DH (2004) Brain 18F-FDG PET in the diagnosis of neurodegenerative dementias: comparison with perfusion SPECT and with clinical evaluations lacking nuclear imaging. J Nucl Med 45:594–607
12. Roberson ED, Mucke L (2006) 100 years and counting: prospects for defeating Alzheimer's disease. Science 314:781–784
13. Selkoe DJ, Hardy J (2016) The amyloid hypothesis of Alzheimer's disease at 25 years. EMBO Mol Med 8:595–608
14. Hardy J (2006) Has the amyloid cascade hypothesis for Alzheimer's disease been proved? Curr Alzheimer Res 3:71–73
15. Goedert M, Spillantini MG (2006) A century of Alzheimer's disease. Science 314:777–781
16. Villemagne VL (2016) Amyloid imaging: past, present and future perspectives. Ageing Res Rev 16:S1568–S1637
17. Scheltens P, Blennow K, Breteler MM, de Strooper B, Frisoni GB, Salloway S et al (2016) Alzheimer's disease. Lancet 388:505–517
18. Vallabhajosula S (2011) Positron emission tomography radiopharmaceuticals for imaging brain Beta-amyloid. Semin Nucl Med 41:283–299
19. Mathis CA, Mason NS, Lopresti BJ, Klunk WE (2012) Development of positron emission tomography beta-amyloid plaque imaging agents. Semin Nucl Med 42:423–432
20. Kung H (2012) The β-amyloid hypothesis in Alzheimer's disease: seeing is believing. ACS Med Chem Lett 3:265–267
21. Kung H, Choi S, Qu W, Zhang W, Skovronsky D (2009) 18F stilbenes and styrylpyridines for PET imaging of A beta plaques in Alzheimer's disease: a miniperspective. J Med Chem 53:933–941
22. Mason NS, Mathis CA, Klunk WE (2013) Positron emission tomography radioligands for in vivo imaging of Abeta plaques. J Labelled Comp Radiopharm 56:89–95
23. Mathis CA, Wang Y, Holt DP, Huang G-F, Debnath ML, Klunk WE (2003) Synthesis and evaluation of ^{11}C-labeled 6-substituted 2-arylbenzothiazoles as amyloid imaging agents. J Med Chem 46:2740–2754
24. Eckelman WC, Mathis CA (2006) Targeting proteins in vivo: in vitro guidelines. Nucl Med Biol 33:161–164
25. Eckelman WC, Mathis CA (2006) Molecular targets. Nucl Med Biol 33:1

26. Eckelman WC, Kilbourn MR, Mathis CA (2009) Specific to nonspecific binding in radiopharmaceutical studies: it's not so simple as it seems. Nucl Med Biol 36:235–237
27. Kung H, Lee C, Zhuang Z, Kung M, Hou C, Plossl K (2001) Novel stilbenes as probes for amyloid plaques. J Am Chem Soc 123:12740–12741
28. Verhoeff N, Wilson A, Takeshita S, Trop L, Hussey D, Singh K et al (2004) In vivo imaging of Alzheimer disease beta-amyloid with [11C]SB-13 PET. Am J Geriatr Psychiatry 12:584–595
29. Ong K, Villemagne VL, Bahar-Fuchs A, Lamb F, Chetelat G, Raniga P et al (2013) 18F-florbetaben Abeta imaging in mild cognitive impairment. Alzheimers Res Ther 5:4
30. Clark CM, Schneider JA, Bedell BJ, Beach TG, Bilker WB, Mintun MA et al (2011) Use of florbetapir-PET for imaging beta-amyloid pathology. JAMA 305:275–283
31. Leinonen V, Rinne JO, Virtanen KA, Eskola O, Rummukainen J, Huttunen J et al (2013) Positron emission tomography with [18F]flutemetamol and [11C]PiB for in vivo detection of cerebral cortical amyloid in normal pressure hydrocephalus patients. Eur J Neurol 20:1043–1052
32. Klunk WE (2011) Amyloid imaging as a biomarker for cerebral beta-amyloidosis and risk prediction for Alzheimer dementia. Neurobiol Aging 32(Suppl 1):S20–S36
33. Villemagne VL, Burnham S, Bourgeat P, Brown B, Ellis KA, Salvado O et al (2013) Amyloid beta deposition, neurodegeneration, and cognitive decline in sporadic Alzheimer's disease: a prospective cohort study. Lancet Neurol 12:357–367
34. Curtis C, Gamez JE, Singh U, Sadowsky CH, Villena T, Sabbagh MN et al (2015) Phase 3 trial of flutemetamol labeled with radioactive fluorine 18 imaging and neuritic plaque density. JAMA Neurol 72:287–294
35. Rowe CC, Pejoska S, Mulligan RS, Jones G, Chan JG, Svensson S et al (2013) Head-to-head comparison of 11C-PiB and 18F-AZD4694 (NAV4694) for beta-amyloid imaging in aging and dementia. J Nucl Med 54:880–886
36. Rowe CC, Jones G, Dore V, Pejoska S, Margison L, Mulligan RS et al (2016) Standardized expression of 18F-NAV4694 and 11C-PiB beta-amyloid PET results with the centiloid scale. J Nucl Med 57:1233–1237
37. Bateman RJ, Xiong C, Benzinger TLS, Fagan AM, Goate A, Fox NC et al (2012) Clinical and biomarker changes in dominantly inherited Alzheimer's disease. N Engl J Med 367:795–804
38. Drzezga A, Grimmer T, Henriksen G, Muhlau M, Perneczky R, Miederer I et al (2009) Effect of APOE genotype on amyloid plaque load and gray matter volume in Alzheimer disease. Neurology 72:1487–1494
39. Tayeb HO, Murray ED, Price BH, Tarazi FI (2013) Bapineuzumab and solanezumab for Alzheimer's disease: is the 'amyloid cascade hypothesis' still alive? Expert Opin Biol Ther 13:1075–1084
40. Grundman M, Dibernardo A, Raghavan N, Krams M, Yuen E (2013) 2012: a watershed year for Alzheimer's disease research. J Nutr Health Aging 17:51–53
41. Callaway E (2012) Alzheimer's drugs take a new tack. Nature 489:13–14
42. Niva C, Parkinson J, Olsson F, van Schaick E, Lundkvist J, Visser SA (2013) Has inhibition of Abeta production adequately been tested as therapeutic approach in mild AD? A model-based meta-analysis of gamma-secretase inhibitor data. Eur J Clin Pharmacol 69:1247–1260
43. Li H, Qin J, Dhondi P, Zhou W, Vicarel M, Bara T et al (2013) The discovery of fused oxadiazepines as gamma secretase modulators for treatment of Alzheimer's disease. Bioorg Med Chem Lett 23:466–471
44. Wu WL, Sasikumar TK, Domalski MS, Qiang L, Burnett DA, Clader J et al (2013) A-ring modification of SCH 900229 and related chromene sulfone gamma-secretase inhibitors. Bioorg Med Chem Lett 23:850–853
45. Huang Y, Li T, Eatherton A, Mitchell WL, Rong N, Ye L et al (2013) Orally bioavailable and brain-penetrant pyridazine and pyridine-derived gamma-secretase modulators reduced amyloidogenic Abeta peptides in vivo. Neuropharmacology 70C:278–286
46. Kakuda N, Akazawa K, Hatsuta H, Murayama S, Ihara Y (2013) Suspected limited efficacy of gamma-secretase modulators. Neurobiol Aging 34:1101–1104

47. Sabri O, Sabbagh MN, Seibyl J, Barthel H, Akatsu H, Ouchi Y et al (2015) Florbetaben PET imaging to detect amyloid beta plaques in Alzheimer's disease: phase 3 study. Alzheimers Dement 11:964–974
48. Brier MR, Gordon B, Friedrichsen K, McCarthy J, Stern A, Christensen J, et al (2016) Tau and Abeta imaging, CSF measures, and cognition in Alzheimer's disease. Sci Transl Med 8:338ra66
49. Holtzman DM, Carrillo MC, Hendrix JA, Bain LJ, Catafau AM, Gault LM, et al (2016) Tau: from research to clinical development. Alzheimers Dement
50. Herholz K (2016) Tau PET, and tauopathies. Eur J Nucl Med Mol Imaging 43:1684–1685
51. Wu JW, Herman M, Liu L, Simoes S, Acker CM, Figueroa H et al (2013) Small misfolded Tau species are internalized via bulk endocytosis and anterogradely and retrogradely transported in neurons. J Biol Chem 288:1856–1870
52. Scholl M, Lockhart SN, Schonhaut DR, O'Neil JP, Janabi M, Ossenkoppele R et al (2016) PET imaging of Tau deposition in the aging human brain. Neuron 89:971–982
53. Attems J, Thal DR, Jellinger KA (2012) The relationship between subcortical tau pathology and Alzheimer's disease. Biochem Soc Trans 40:711–715
54. Mandelkow EM, Mandelkow E (2012) Biochemistry and cell biology of tau protein in neurofibrillary degeneration. Cold Spring Harb Perspect Med 2, a006247
55. Spillantini MG, Goedert M (2013) Tau pathology and neurodegeneration. Lancet Neurol 12:609–622
56. Daebel V, Chinnathambi S, Biernat J, Schwalbe M, Habenstein B, Loquet A et al (2012) beta-Sheet core of tau paired helical filaments revealed by solid-state NMR. J Am Chem Soc 134:13982–13989
57. Holtzman DM, Carrillo MC, Hendrix JA, Bain LJ, Catafau AM, Gault LM et al (2016) Tau: from research to clinical development. Alzheimers Dement 16:S1552–S5260
58. Sala Frigerio C, De Strooper B (2016) Alzheimer's disease mechanisms and emerging roads to novel therapeutics. Annu Rev Neurosci 39:57–79
59. Villemagne VL, Okamura N (2016) Tau imaging in the study of ageing, Alzheimer's disease, and other neurodegenerative conditions. Curr Opin Neurobiol 36:43–51
60. Small GW, Siddarth P, Kepe V, Ercoli LM, Burggren AC, Bookheimer SY et al (2012) Prediction of cognitive decline by positron emission tomography of brain amyloid and tau. Arch Neurol 69:215–222
61. Shoghi-Jadid K, Small GW, Agdeppa ED, Kepe V, Ercoli LM, Siddarth P et al (2002) Localization of neurofibrillary tangles and beta-amyloid plaques in the brains of living patients with Alzheimer disease: binding characteristics of radiofluorinated 6-dialkylamino-2-naphthylethylidene derivatives as positron emission tomography imaging probes for beta-amyloid plaques in Alzheimer disease. Am J Geriatr Psychiatry 10:24–35
62. Agdeppa ED, Kepe V, Liu J, Flores-Torres S, Satyamurthy N, Petric A et al (2001) Binding characteristics of radiofluorinated 6-dialkylamino-2-naphthylethylidene derivatives as positron emission tomography imaging probes for b-amyloid plaques in Alzheimer's disease. J Neurosci 21:RC189
63. Small GW, Kepe V, Siddarth P, Ercoli LM, Merrill DA, Donoghue N et al (2013) PET scanning of brain tau in retired national football league players: preliminary findings. Am J Geriatr Psychiatry 21:138–144
64. Kepe V, Bordelon Y, Boxer A, Huang SC, Liu J, Thiede FC et al (2013) PET imaging of neuropathology in tauopathies: progressive supranuclear palsy. J Alzheimers Dis 36:145–153
65. Declercq L, Celen S, Lecina J, Ahamed M, Tousseyn T, Moechars D et al (2016) Comparison of new Tau PET-tracer candidates with [18F]T808 and [18F]T807. Mol Imaging 15:36012115624920
66. Chien DT, Bahri S, Szardenings AK, Walsh JC, Mu F, Su MY et al (2013) Early clinical PET imaging results with the novel PHF-tau radioligand [F-18]-T807. J Alzheimers Dis 34:457–468

67. Zhang W, Arteaga J, Cashion DK, Chen G, Gangadharmath U, Gomez LF et al (2012) A highly selective and specific PET tracer for imaging of Tau pathologies. J Alzheimers Dis 31:601–612
68. Johnson KA, Schultz A, Betensky RA, Becker JA, Sepulcre J, Rentz D et al (2016) Tau positron emission tomographic imaging in aging and early Alzheimer disease. Ann Neurol 79:110–119
69. Dani M, Brooks DJ, Edison P (2015) Tau imaging in neurodegenerative diseases. Eur J Nucl Med Mol Imaging 43:1139–1150
70. Kimura Y, Endo H, Ichise M, Shimada H, Seki C, Ikoma Y et al (2016) A new method to quantify tau pathologies with (11)C-PBB3 PET using reference tissue voxels extracted from brain cortical gray matter. Eur J Nucl Med Mol Imaging Res 6:24
71. Xia CF, Arteaga J, Chen G, Gangadharmath U, Gomez LF, Kasi D et al (2013) [(18)F]T807, a novel tau positron emission tomography imaging agent for Alzheimer's disease. Alzheimers Dement 9:666–676
72. Marquie M, Normandin MD, Vanderburg CR, Costantino IM, Bien EA, Rycyna LG et al (2015) Validating novel tau positron emission tomography tracer [F-18]-AV-1451 (T807) on postmortem brain tissue. Ann Neurol 78:787–800
73. Okamura N, Suemoto T, Furumoto S, Suzuki M, Shimadzu H, Akatsu H et al (2005) Quinoline and benzimidazole derivatives: candidate probes for in vivo imaging of tau pathology in Alzheimer's disease. J Neurosci 25:10857–10862
74. Shimadzu H, Suemoto T, Suzuki M, Shiomitsu T, Okamura N, Kudo Y et al (2004) Novel probes for imaging amyloid-b: F-18 and C-11 labeling of 2-(4-aminostyryl)benzoxazole derivatives. J Labelled Compd Radiopharm 47:181–190
75. Okamura N, Suemoto T, Shimadzu H, Suzuki M, Shiomitsu T, Akatsu H et al (2004) Styrylbenzoxazole derivatives for in vivo imaging of amyloid plaques in the brain. J Neurosci 24:2535–2541
76. Kimura Y, Ichise M, Ito H, Shimada H, Ikoma Y, Seki C et al (2015) PET quantification of tau pathology in human brain with 11C-PBB3. J Nucl Med 56:1359–1365
77. Wang M, Gao M, Xu Z, Zheng QH (2015) Synthesis of a PET tau tracer [(11)C]PBB3 for imaging of Alzheimer's disease. Bioorg Med Chem Lett 25:4587–4592
78. Hashimoto H, Kawamura K, Takei M, Igarashi N, Fujishiro T, Shiomi S et al (2015) Identification of a major radiometabolite of [11C]PBB3. Nucl Med Biol 42:905–910
79. Chiotis K, Saint-Aubert L, Savitcheva I, Jelic V, Andersen P, Jonasson M et al (2016) Imaging in-vivo tau pathology in Alzheimer's disease with THK5317 PET in a multimodal paradigm. Eur J Nucl Med Mol Imaging 43:1686–1699
80. Harada R, Okamura N, Furumoto S, Furukawa K, Ishiki A, Tomita N et al (2016) 18F-THK5351: a novel PET radiotracer for imaging neurofibrillary pathology in Alzheimer disease. J Nucl Med 57:208–214
81. Rodriguez-Vieitez E, Leuzy A, Chiotis K, Saint-Aubert L, Wall A, Nordberg A (2016) Comparability of [18F]THK5317 and [11C]PIB blood flow proxy images with [18F]FDG positron emission tomography in Alzheimer's disease. J Cereb Blood Flow Metab (Epub ahead of print)
82. Tago T, Furumoto S, Okamura N, Harada R, Adachi H, Ishikawa Y et al (2016) Structure-activity relationship of 2-arylquinolines as PET imaging tracers for tau pathology in Alzheimer disease. J Nucl Med 57:608–614
83. Harada R, Okamura N, Furumoto S, Tago T, Maruyama M, Higuchi M et al (2013) Comparison of the binding characteristics of [18F]THK-523 and other amyloid imaging tracers to Alzheimer's disease pathology. Eur J Nucl Med Mol Imaging 40:125–132
84. Okamura N, Harada R, Furukawa K, Furumoto S, Tago T, Yanai K et al (2016) Advances in the development of tau PET radiotracers and their clinical applications. Ageing Res Rev 30:107–113

85. Okamura N, Furumoto S, Harada R, Tago T, Yoshikawa T, Fodero-Tavoletti M et al (2013) Novel 18F-labeled arylquinoline derivatives for noninvasive imaging of tau pathology in Alzheimer disease. J Nucl Med 54:1420–1427
86. Rojo LE, Alzate-Morales J, Saavedra IN, Davies P, Maccioni RB (2010) Selective interaction of lansoprazole and astemizole with tau polymers: potential new clinical use in diagnosis of Alzheimer's disease. J Alzheimers Dis 19:573–589
87. Fawaz MV, Brooks AF, Rodnick ME, Carpenter GM, Shao X, Desmond TJ et al (2014) High affinity radiopharmaceuticals based upon lansoprazole for PET imaging of aggregated tau in Alzheimer's disease and progressive supranuclear palsy: synthesis, preclinical evaluation, and lead selection. ACS Chem Neurosci 5:718–730
88. Shao X, Carpenter GM, Desmond TJ, Sherman P, Quesada CA, Fawaz M et al (2012) Evaluation of [(11)C]N-methyl lansoprazole as a radiopharmaceutical for PET imaging of tau neurofibrillary tangles. ACS Med Chem Lett 3:936–941
89. Walji AM, Hostetler ED, Selnick H, Zeng Z, Miller P, Bennacef I et al (2016) Discovery of 6-(Fluoro-(18)F)-3-(1H-pyrrolo[2,3-c]pyridin-1-yl)isoquinolin-5-amine ([(18)F]-MK-6240): a positron emission tomography (PET) imaging agent for quantification of neurofibrillary tangles (NFTs). J Med Chem 59:4778–4789
90. Cai L, Qu B, Hurtle BT, Dadiboyena S, Diaz-Arrastia R, Pike VW (2016) Candidate PET radioligand development for neurofibrillary tangles: two distinct radioligand binding sites identified in postmortem Alzheimer's disease brain. ACS Chem Neurosci 7:897–911

Index